INTRODUCTORY COMPLEX ANALYSIS AND APPLICATIONS

*INTRODUCTORY
COMPLEX
AND*

ANALYSIS
APPLICATIONS

William R. Derrick

UNIVERSITY OF UTAH

ACADEMIC PRESS   NEW YORK AND LONDON

COPYRIGHT © 1972, BY ACADEMIC PRESS, INC.
ALL RIGHTS RESERVED
NO PART OF THIS BOOK MAY BE REPRODUCED IN ANY FORM,
BY PHOTOSTAT, MICROFILM, RETRIEVAL SYSTEM, OR ANY
OTHER MEANS, WITHOUT WRITTEN PERMISSION FROM
THE PUBLISHERS.

ACADEMIC PRESS, INC.
111 Fifth Avenue, New York, New York 10003

United Kingdom Edition published by
ACADEMIC PRESS, INC. (LONDON) LTD.
24/28 Oval Road, London NW1 7DD

LIBRARY OF CONGRESS CATALOG CARD NUMBER: 70 - 185029

AMS (MOS) 1970 Subject Classification: 30-01

PRINTED IN THE UNITED STATES OF AMERICA

CONTENTS

Preface ix
Table of Symbols xi

1. ANALYTIC FUNCTIONS

1.1. Complex Numbers 1
1.2. Properties of the Complex Plane 7
1.3. Functions of a Complex Variable 10
1.4. Sufficient Conditions for Analyticity 15
1.5. Some Elementary Functions 18
1.6. Continuation 21
   Notes 27

2. COMPLEX INTEGRATION

2.1.	Line Integrals	28
2.2.	The Cauchy–Goursat Theorem	33
2.3.	The Fundamental Theorem of Integration	38
2.4.	The Cauchy Integral Formula	46
2.5.	Liouville's Theorem and the Maximum Principle	54
	Notes	57

3. INFINITE SERIES

3.1.	Taylor Series	58
3.2.	Uniform Convergence of Series	64
3.3.	Laurent Series	71
3.4.	Isolated Singularities	76
3.5.	Analytic Continuation	80
3.6.	Riemann Surfaces	84
	Notes	88

4. CONTOUR INTEGRATION

4.1.	The Residue Theorem	89
4.2.	Evaluation of Improper Real Integrals	94
4.3.	Continuation	98
4.4.	Integration of Multivalued Functions	103
4.5.	Other Integration Techniques	107
4.6.	The Argument Principle	111
	Notes	115

5. CONFORMAL MAPPINGS

5.1.	General Properties	116
5.2.	Linear Fractional Transformations	119
5.3.	Continuation	124
5.4.	The Schwarz–Christoffel Formula	128
5.5.	Physical Applications	134
	Notes	144

6. BOUNDARY-VALUE PROBLEMS

6.1.	Harmonic Functions	145
6.2.	Poisson's Integral Formula	148

6.3.	Applications	152
	Notes	162

7. FOURIER AND LAPLACE TRANSFORMATIONS

7.1.	Fourier Series	163
7.2.	Fourier Transforms	169
7.3.	Laplace Transforms	174
7.4.	Properties of Laplace Transforms	183
	Notes	190

8. ASYMPTOTIC EXPANSIONS

8.1.	Definitions and Properties	191
8.2.	Method of Steepest Descent	197
8.3.	Continuation	202
	Notes	207

APPENDIX

A.1.	Table of Conformal Mappings	209

References 213
Index 215

PREFACE

My objective has been to provide a compact, introductory treatment of the functions of a complex variable, emphasizing applications, in a manner suitable for use in a one-semester course for undergraduate students in engineering and mathematics. To this end, I have moved rapidly through the preliminaries by emphasizing the similarities with real variables and elaborating only on the differences between the two theories. It should therefore be possible to begin the study of complex integration in the third week, and the time gained can be used for a richer development of the theory and its applications.

To avoid making the presentation too theoretical for engineers or too applied for mathematicians, those sections containing long proofs also include extended practical applications of the theory. By omitting either proof or application the instructor can tailor his course to the abilities and interests of his students. None of the proofs presented utilizes material more advanced than is covered in the first semester of an advanced calculus course or, indeed, in many of the newer elementary calculus textbooks. The exercises at the end of each section are ordered by level of difficulty, the odd problems complementing the even ones in each set. Throughout the text, optional sections are indicated by a dagger (†).

I have included two topics which are not usually found in textbooks at this level: Integral transforms and asymptotic expansions. These subjects are usually studied in a real variable setting, but assume far greater significance and power when viewed as complex variables. Indeed, the Poisson integral formula can seldom be solved without using transform methods. These sections include proofs of the inversion theorems, some residue solutions of inverse Laplace transforms, and a short proof of Stirling's Formula by the method of steepest descent.

The notes at the end of each chapter indicate directions for further generalization and study. Moreover, they serve to indicate a few of the many topics that were omitted with regret as being too specialized for a work of this nature.

The material presented approximates the recommendations of the Committee on the Undergraduate Program in Mathematics (CUPM), and able, well prepared students could conceivably cover the entire book in a semester,

I would like to thank Joyce Kiser for the excellent work she did in typing the manuscript, and my wife, Judith, for her patience.

TABLE OF SYMBOLS

The number indicates the page on which the symbol is defined.

$\mathscr{C}$	2	$\sin z$	24	V	134
i	2	$\cosh z$	25	V_n	135
z	2	$\sinh z$	25	V_s	135
$|z|$	2	pwd	29	Δ	145
$\arg z$	2	$-\gamma$	29	$\Gamma + iQ$	155
$\text{Arg } z$	2	$|dz|$	30	$U(\phi + 0)$	166
$\text{Re } z$	2	$\int_\gamma f(z)\, dz$	30	$U(\phi - 0)$	166
$\text{Im } z$	2	∂R	34	$U'(\phi + 0)$	166
$\bar{z}$	4	$P_n(z)$	53	$U'(\phi - 0)$	166
$\mathscr{M}$	5	$L_n(z)$	53	$\hat{U}$	171
∞	5	S_n	58	$H(\phi - a)$	174
Int S	8	lub	66	$\mathscr{L}_2\{U\}$	175
$f(z)$	10	$J_n(z)$	76	$\mathscr{L}\{U\}$	175
e^z	19	(f, G)	80	$\text{Si}(\phi)$	182
$\mathscr{R}$	20	$\Gamma(z)$	84	$\delta(\phi - a)$	182
$\log z$	22	$\mathscr{F}$	85	$U * V$	185
$\text{Log } z$	22	a_{-1}	89	$Z(s)$	188
z^a	22	$\text{Res}_z f(z)$	89	$f \sim g$	192
$\cos z$	24	PV	99	$\text{Ei}(z)$	192

INTRODUCTORY COMPLEX ANALYSIS AND APPLICATIONS

Chapter 1 ANALYTIC FUNCTIONS

The theory of functions of a complex variable extends the concepts of calculus to the complex plane. In so doing, differentiation and integration acquire new depth and elegance, and the two-dimensional nature of the complex plane yields many results useful in applied mathematics.

1.1 COMPLEX NUMBERS

The numbers used in elementary algebra and calculus represented by the points on a straight line are called *real numbers*. The impossibility of solving equations such as $x^2 + 1 = 0$ by real numbers led to the creation of the *complex numbers*. Although complex numbers are now universally accepted, their development occasioned much opposition, which finally vanished because of their usefulness and the unity they brought to many topics in mathematics. The complex number system $\mathscr{C}$ consists of all numbers of the form

$$z = x + iy,$$

where x and y are real numbers and i is the *imaginary unit* satisfying the property $i^2 = -1$. The number x is called the *real* part of z and denoted by Re z; the number y is called the *imaginary* part of z and denoted by Im z. If $x = 0$, we have $z = iy$ and say that z is *pure imaginary*.

As a model for the complex number system we use the *complex plane*. The number $x + iy$ is depicted geometrically as the point with the coordinates (x, y) in the usual Cartesian plane. The x-axis is called the *real axis* and the y-axis is referred to as the *imaginary axis*. The *origin* of the coordinate system corresponds to the complex number 0 (see Figure 1.1). We may also use polar

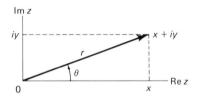

Figure 1.1
The complex plane.

coordinates in the complex plane to obtain a representation for the number $z = x + iy$ ($\neq 0$). Since $x = r \cos \theta$, $y = r \sin \theta$ we have

$$z = r(\cos \theta + i \sin \theta), \qquad z \neq 0.$$

The positive number $r = \sqrt{x^2 + y^2}$ is the distance from z to the origin; we call it the *modulus* or *absolute value* of z and denote it by the symbol $|z|$. The angle $\theta = \tan^{-1} y/x$, determined except for a multiple of 2π, is called the *argument* of z and denoted by arg z; that value of arg z satisfying

$$-\pi \leq \arg z < \pi$$

is called the *principal value* of the argument and is designated by Arg z. When working with the argument, it is convenient to adopt the convention that the

1.1 COMPLEX NUMBERS

notation arg z ignores multiples of 2π, and to use the expression

$$\text{Arg } z + 2\pi k, \qquad k \text{ a fixed integer,}$$

to indicate a particular angle. Thus

$$z = |z|[\cos(\arg z) + i \sin(\arg z)], \qquad z \neq 0.$$

It is common practice in engineering books to refer to the *magnitude* and *angle* instead of the absolute value and argument of a complex number, as well as to use the letter j for the imaginary unit.

Addition of complex numbers is performed by adding the real parts together to obtain the real part of the sum, and doing the same for the imaginary parts

$$(x + iy) + (a + ib) = (x + a) + i(y + b).$$

Geometrically this amounts to vector addition in the plane (see Figure 1.2).

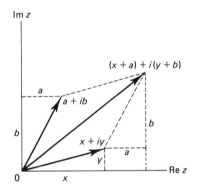

Figure 1.2

Vector addition.

Multiplication is performed by a formal application of the distributive law together with the relation $i^2 = -1$

$$(x + iy)(a + ib) = xa + i(ya + xb) + i^2 yb = (xa - yb) + i(ya + xb).$$

Division is accomplished as

$$\frac{a + ib}{x + iy} = \frac{a + ib}{x + iy} \cdot \frac{x - iy}{x - iy} = \frac{ax + by}{x^2 + y^2} + i\frac{bx - ay}{x^2 + y^2}.$$

The geometrical significance of these last two operations may be obtained by introducing polar coordinates. Let

$$z = x + iy = r(\cos\theta + i\sin\theta), \qquad w = a + ib = \rho(\cos\varphi + i\sin\varphi),$$

then using the addition formulas of trigonometry

$$zw = (x + iy)(a + ib) = r\rho[\cos(\theta + \varphi) + i\sin(\theta + \varphi)],$$

and

$$\frac{w}{z} = \frac{a+ib}{x+iy} = \frac{\rho}{r}[\cos(\varphi - \theta) + i\sin(\varphi - \theta)].$$

For multiplication the angle between w and zw must be identical to the angle between 1 and z in Figure 1.3. It follows that the triangles 0, 1, z and 0, w, zw

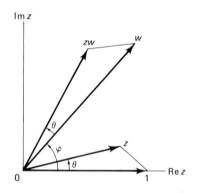

Figure 1.3
Complex multiplication.

are similar. The geometric construction for division is the same, except the similar triangles are now 0, 1, z and 0, w/z, w.

We call $x - iy$ the *complex conjugate* of $z = x + iy$ and denote it by $\bar{z}$. Observe that $z\bar{z} = |z|^2$, $z + \bar{z} = 2\operatorname{Re} z$, and $z - \bar{z} = 2i\operatorname{Im} z$. Since $\overline{(z_1 z_2)} = \bar{z}_1 \cdot \bar{z}_2$, it follows that

$$|z_1 z_2|^2 = (z_1 z_2)\overline{(z_1 z_2)} = z_1 \bar{z}_1 z_2 \bar{z}_2 = |z_1|^2 |z_2|^2,$$

hence

$$|z_1 z_2| = |z_1| \, |z_2|.$$

The complex conjugate of z is often denoted by z^* in engineering books.
We now prove:

1.1 COMPLEX NUMBERS

The Triangle Inequality

$$|z_1 + z_2| \leq |z_1| + |z_2|.$$

Proof The distance between z_1 and $-z_2$ is given by

$$\sqrt{(x_1 + x_2)^2 + (y_1 + y_2)^2} = |(x_1 + x_2) + i(y_1 + y_2)| = |z_1 + z_2|.$$

From elementary geometry we know that in the triangle with vertices 0, z_1, $-z_2$, the length of one side is no bigger than the sum of the lengths of the other two sides, that is,

$$|z_1 + z_2| \leq |z_1| + |z_2|$$

(see Figure 1.4).

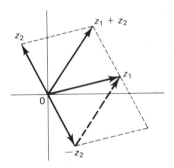

Figure 1.4
The Triangle Inequality.

For many purposes it is useful to extend the system $\mathscr{C}$ of complex numbers by introducing the symbol ∞, the *point at infinity*, into the plane. This new set is called the *extended complex plane* $\mathscr{M}$ and the point ∞ satisfies the following algebraic rules.

$$a + \infty = \infty + a = \infty, \qquad \frac{a}{\infty} = 0 \qquad \text{for} \quad a \neq \infty,$$

$$b \cdot \infty = \infty \cdot b = \infty, \qquad \frac{b}{0} = \infty \qquad \text{for} \quad b \neq 0.$$

As a geometrical model for $\mathscr{M}$ we use the unit sphere $x_1^2 + x_2^2 + x_3^2 = 1$ in three-dimensional space. We associate to each point z in the plane, that point Z in the sphere where the ray originating from the north pole N and passing through z intersects the sphere. Thus, N corresponds to ∞ (see Figure 1.5). This model is called the *Riemann sphere* and the point correspondence is referred to as *stereographic projection*. Observe that all straight lines in $\mathscr{C}$ correspond to circles passing through ∞ in $\mathscr{M}$.

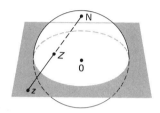

Figure 1.5
The Riemann sphere.

EXERCISES

1. Find the modulus, argument, and polar form of the following complex numbers:

   (a) i,
   (b) $1 + i$,
   (c) $-3 + 4i$,
   (d) $5 - 12i$.

2. Find the sum, product, and quotient of the following pairs of complex numbers:

   (a) $1 + i$, $1 - i$,
   (b) $2 + i$, $3 - 4i$.

3. Prove that:

   (a) $\overline{(z_1 \pm z_2)} = \overline{z_1} \pm \overline{z_2}$,
   (b) $\overline{(z_1 z_2)} = \overline{z_1} \cdot \overline{z_2}$,
   (c) $\overline{\left(\dfrac{z_1}{z_2}\right)} = \dfrac{\overline{z_1}}{\overline{z_2}}$, $z_2 \neq 0$,
   (d) $|z| = |\overline{z}|$.

4. Prove that:

   (a) $\arg(\overline{z}) = -\arg z$,
   (b) $\arg(z_1 z_2) = \arg z_1 + \arg z_2$,
   (c) $\arg \dfrac{z_1}{z_2} = \arg z_1 - \arg z_2$.

1.2 PROPERTIES OF THE COMPLEX PLANE

5. Prove that:
 (a) $|z_1 - z_2| \geq ||z_1| - |z_2||$,
 (b) $\left|\sum_{k=1}^{n} z_k\right| \leq \sum_{k=1}^{n} |z_k|$,
 (c) $|z_1 \pm z_2|^2 = |z_1|^2 + |z_2|^2 \pm 2\,\text{Re}\,z_1\bar{z}_2$,
 (d) $|\text{Re}\,z| + |\text{Im}\,z| \leq \sqrt{2}|z|$.

6. Prove that
$$\left|\frac{z-a}{1-\bar{a}z}\right| < 1$$
if $|z| < 1$ and $|a| < 1$.

7. Let $z = \cos\theta + i\sin\theta$. Show $\arg(z^n) = n\arg z$, and then obtain De Moivre's Theorem
$$(\cos\theta + i\sin\theta)^n = \cos n\theta + i\sin n\theta.$$

8. By minimizing the expression $\sum_{k=1}^{n}(|a_k| - \lambda|z_k|)^2$, where $a_1, \ldots, a_n, z_1, \ldots, z_n$ are complex numbers, for arbitrary real λ, show that
$$\left(\sum_{k=1}^{n}|a_k z_k|\right)^2 \leq \left(\sum_{k=1}^{n}|a_k|^2\right)\left(\sum_{k=1}^{n}|z_k|^2\right).$$

1.2 PROPERTIES OF THE COMPLEX PLANE

Let z_0 be a complex number. Then an *ε-neighborhood* of z_0 is the set of all points z whose distance from z_0 is less than ε, that is all z satisfying $|z - z_0| < \varepsilon$ (see Figure 1.6). Pictorially, this is the "inside" of a disk centered at z_0 of radius ε.

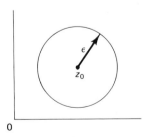

Figure 1.6
An ε-neighborhood of z_0.

Let S be a set of points in the complex plane $\mathscr{C}$. The *complement* of S is the set $\mathscr{C} - S$ of all points not in S. The point z_0 is said to be an *interior point* of S if some ε-neighborhood of z_0 is contained entirely in S; z_0 is an *exterior point* of S if some ε-neighborhood of z_0 is contained entirely in $\mathscr{C} - S$; otherwise z_0 is said to be a *boundary point* of S. The set of all boundary points of S is called the *boundary* of S, and the set of all interior points is referred to as the *interior* of S and denoted by Int S. A set S is *open* if all its points are interior points, that is, $S = $ Int S (see Figure 1.7). The complement of an open set is said to be *closed*. For example, the set S' of all points z such that $|z| < 1$, is open and the set S'' of all points z such that $|z| \leq 1$, is closed.

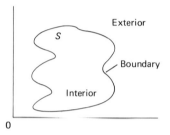

Figure 1.7

A set S is said to be *bounded* if there is a positive real number R such that all z in S satisfy $|z| < R$. If this condition does not hold, we say that S is *unbounded*. As examples, the S'' above is bounded but the first quadrant is unbounded. A set S is *connected* if it cannot be represented as the union of two nonempty disjoint sets A and B neither containing a boundary point of the other. Intuitively, what this says is that S consists of a single piece. For example S' is connected, but the set of all z for which $|z - 2| < 1$ or $|z + 2| < 1$ is not connected, as we can let A be the set of all z such that $|z - 2| < 1$, and B be the set of all z for which $|z + 2| < 1$ (see Figure 1.8).

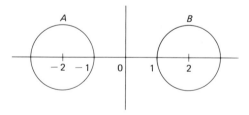

Figure 1.8

1.2 PROPERTIES OF THE COMPLEX PLANE

A *domain* is an open connected set. It is intuitively clear that any two points in a domain can be joined by a polygon contained in the domain, but this fact requires verification. The proof is slightly complicated, but should be studied as the technique will be used again.

Theorem Any two points of a domain can be joined by a polygon which lies in the domain.

Proof† Call the domain S, and suppose z_0 lies in S. Denote by S_1 all those points in S which can be joined to z_0 by a polygon and by S_2 those that cannot be so joined. If z_1 is in S_1, and hence in S, it is an interior point of S thus there is an ε-neighborhood of z_1 lying in S: $|z - z_1| < \varepsilon$. All these points are in S_1 as each can be joined to z_1 by a straight line lying in S and hence can be joined to z_0 by a polygon in S. Thus every point in S_1 is an interior point of S_1 and so S_1 is open. If z_2 is in S_2, let $|z - z_2| < \varepsilon$ be a neighborhood contained in S. No point in this neighborhood can be in S_1 for otherwise z_2 is in S_1. Thus, every point of S_2 is an interior point of S_2, so S_2 is open. Neither set can contain a boundary point of the other as both are open. Since S is connected one of these sets must be empty, but z_0 is in S_1, so S_2 is empty. Thus any two points can be joined to z_0 by a polygonal path in S and thus to each other by a polygonal path by way of z_0. The proof is now complete.

Furthermore, it is even possible to insist that all lines in the polygon be parallel to the coordinate axes. The proof using this additional requirement is identical as we can always join the center of an open disk to one of its points by at most two line segments parallel to the axes.

A domain is *simply connected* if its complement in $\mathcal{M}$ is connected. This merely says that a simply connected domain has no "holes" in it. For example S' is simply connected, but the set of all z satisfying $0 < |z| < 1$ is not.

Elementary geometric figures in the complex plane, such as ellipses, parabolas, and so on, may be defined in terms of distances. For example, the ellipse with foci at ± 1 and major axis of length 3 is given by the equation

$$|z + 1| + |z - 1| = 3,$$

and its "inside" is given by

$$|z + 1| + |z - 1| < 3.$$

EXERCISES

1. Classify the following sets according to the terms open, closed, bounded, connected, domain, simply connected:
   (a) $|z + 3| < 2$,
   (b) $|\text{Re } z| < 1$,
   (c) $|\text{Im } z| > 1$,
   (d) $0 < |z - 1| \leq 1$,
   (e) $|z| \leq \text{Re } z + 2$,
   (f) $|z - 1| - |z + 1| > 2$,
   (g) $\text{Im } \dfrac{z - 1}{z - 2} = 0$,
   (h) $\text{Re } \dfrac{z - 1}{z - 2} \leq 0$.

2. What are the boundaries of the sets in Exercise 1?
3. Write the general equation in complex form of a hyperbola with foci at a and b.
4. What is the equation in complex form for the parabola $y = x^2 - \frac{1}{4}$?
5. Prove the following properties of open and closed sets:
   (a) The intersection of finitely many open sets is open;
   (b) The union of finitely many closed sets is closed;
   (c) The intersection of any collection of closed sets is closed;
   (d) The union of any collection of open sets is open.
6. Prove that if any two points in an open set can be joined by a polygon lying in the set, then the set is a domain.
7. The *closure* of a set S is the intersection of all closed sets containing S. Prove that the closure of a connected set is connected.

1.3 FUNCTIONS OF A COMPLEX VARIABLE

A complex-valued function of a complex variable is a rule which assigns to each complex number z in a set S a complex number w. We write $w = f(z)$ and say that w is the value of the function at the point z.

Real-valued functions of a real variable are defined in the same way and the function $y = f(x)$ is exhibited geometrically by a graph in the xy-plane. Unfortunately no such convenient graphical representation is possible for $w = f(z)$, as it would require four dimensions. Instead, information about the function is displayed by drawing separate complex planes for the variables z and w and indicating correspondences between points or sets of points in the two planes (see Figure 1.9). The function f is said to be a *mapping* of the set S in the z-plane into the w-plane. A function f mapping a set S into a set S', $f: S \to S'$, is said to be *one-to-one* if $f(z_1) = f(z_2)$ only for $z_1 = z_2$; it is said to be *onto* if $S' = f(S)$. We call $f(S)$ the *image* set of S under the mapping f.

1.3 FUNCTIONS OF A COMPLEX VARIABLE

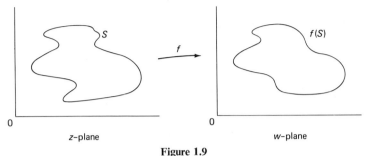

Figure 1.9

A mapping.

Suppose f is defined on a domain G and a is a point in G. Then limits and continuity are defined in the same way as in the real case:

Definition The function $f(z)$ is said to have *limit* A as z approaches a,

$$\lim_{z \to a} f(z) = A,$$

provided that for every $\varepsilon > 0$ there exists a number $\delta > 0$ such that

$$|f(z) - A| < \varepsilon,$$

whenever $0 < |z - a| < \delta$. The function $f(z)$ is said to be *continuous* at a if and only if

$$\lim_{z \to a} f(z) = f(a)$$

(see Figure 10). A continuous function is one that is continuous at all points where it is defined.

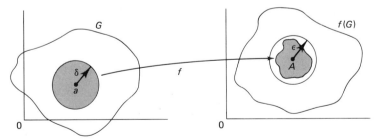

Figure 1.10

Continuity of f at a.

The usual results concerning the limit of a sum, product, and quotient are proved as in the real case, as are the variants where a or A is infinite (see Exercise 1). The only difference is that in the real case we distinguish between $+\infty$ and $-\infty$, whereas in the complex case there is only one infinite limit. The sum $f(z) + g(z)$ and product $f(z) \cdot g(z)$ of two continuous functions are continuous, and the quotient $f(z)/g(z)$ is defined and continuous at a provided $g(a) \neq 0$. If $f(z)$ is continuous, so are $\operatorname{Re} f(z)$, $\operatorname{Im} f(z)$, and $|f(z)|$.

The derivative of a function also have the same formal definition as that given in the real case:

Definition The *derivative f'* of f at a is defined by

$$f'(a) = \lim_{h \to 0} \frac{f(a+h) - f(a)}{h},$$

provided the limit exists. The function f is said to be *analytic* on the domain G if it has a derivative at each point of G, and f is said to be *entire* if it is analytic on all of $\mathscr{C}$.

Observe that in the definition above h is a complex number, as is the quotient $[f(a+h) - f(a)]/h$. Thus, in order for the derivative to exist, it is necessary that the quotient above tend to a unique complex number $f'(a)$ independent of the manner in which h approaches zero.

Formal manipulations of the definition of derivative lead to the usual rules of differentiation:

$$(f \pm g)' = f' \pm g', \qquad (fg)' = fg' + gf',$$

$$\left(\frac{f}{g}\right)' = \frac{gf' - fg'}{g^2}, \quad g \neq 0, \quad (f(g(z)))' = f'(g(z))g'(z), \quad \text{chain rule.}$$

As an example, let $f(z) = z^n$ with n a positive integer. Using the Binomial Theorem we have

$$f'(z) = \lim_{h \to 0} \frac{f(z+h) - f(z)}{h} = \lim_{h \to 0} \frac{(z+h)^n - z^n}{h} = nz^{n-1}.$$

In particular, it follows that every polynomial

$$P(z) = a_0 + a_1 z + a_2 z^2 + \cdots + a_n z^n$$

is entire, since at each point z in $\mathscr{C}$ it has the derivative

$$P'(z) = a_1 + 2a_2 z + \cdots + na_n z^{n-1}.$$

1.3 FUNCTIONS OF A COMPLEX VARIABLE

Nevertheless, there is a fundamental difference between differentiation for functions of a real and of a complex variable. Let $z = (x, y)$ and suppose h is real, then

$$f'(z) = \lim_{h \to 0} \frac{f(x + h, y) - f(x, y)}{h} = f_x(z),$$

and, if $h = ik$ is purely imaginary,

$$f'(z) = \lim_{k \to 0} \frac{f(x, y + k) - f(x, y)}{ik} = \frac{1}{i} f_y(z) = -i f_y(z).$$

Thus, the existence of a complex derivative forces the function to satisfy the partial differential equation

$$f_x = -i f_y.$$

Writing $f(z) = u(z) + iv(z)$, where u and v are real-valued functions of a complex variable, we have

$$u_x + i v_x = f_x = -i f_y = v_y - i u_y,$$

yielding the equations

$$u_x = v_y, \qquad v_x = -u_y.$$

These are the *Cauchy–Riemann* differential equations. We have proved:

Theorem If the function $f(z) = u(z) + iv(z)$ has a derivative at the point z, then the first partial derivatives of u and v, with respect to x and y, exist and satisfy the Cauchy–Riemann equations.

To illustrate let $f(z) = z^2 = (x^2 - y^2) + 2xyi$. Since f is entire, $u = x^2 - y^2$ and $v = 2xy$ must satisfy the Cauchy–Riemann equations

$$u_x = 2x = v_y, \qquad -u_y = +2y = v_x.$$

On the other hand, if $f(z) = |z|^2 = x^2 + y^2$, then $u = x^2 + y^2$, $v = 0$ and $u_x = 2x$, $u_y = 2y$, $v_x = 0 = v_y$, so f satisfies the Cauchy–Riemann equations only at 0, and thus is differentiable only when $z = 0$, since

$$f'(0) = \lim_{h \to 0} \frac{|h|^2}{h} = 0.$$

EXERCISES

1. Suppose $\lim_{z \to a} f(z) = A$, $\lim_{z \to a} g(z) = B$ with f and g defined on some domain G containing the point a. Prove for finite a, A, B:
   (a) $\lim\limits_{z \to a} f(z) \pm g(z) = A \pm B$,
   (b) $\lim\limits_{z \to a} f(z)g(z) = AB$,
   (c) $\lim\limits_{z \to a} \dfrac{f(z)}{g(z)} = \dfrac{A}{B}$, provided $B \neq 0$.

2. Let f and g be continuous functions defined on a domain G. Prove the following functions are continuous on G:
   (a) $f(z) + g(z)$,
   (b) $f(z)g(z)$,
   (c) $\dfrac{f(z)}{g(z)}$, provided $g(z) \neq 0$ for any z in G,
   (d) $\operatorname{Re} f(z)$,
   (e) $|f(z)|$.

3. Let f and g be analytic functions defined on the domain G. Prove the following rules of differentiation:
   (a) $(f \pm g)' = f' \pm g'$,
   (b) $(fg)' = fg' + gf'$,
   (c) $\left(\dfrac{f}{g}\right)' = \dfrac{gf' - fg'}{g^2}$, provided $g(z) \neq 0$ for any z in G.

4. Prove that if f is differentiable at the point a, then it is continuous at a.
5. Show that the quotient $P(z)/Q(z)$ of two polynomials has a derivative at every point z where $Q(z) \neq 0$.
6. Using the rules for differentiation find the derivatives of the following functions:
   (a) $f(z) = 18z^3 - \dfrac{z^2}{4} + 4z + 8$,
   (b) $f(z) = (2z^3 + 1)^5$,
   (c) $f(z) = \dfrac{z+1}{z-1}$, $\quad z \neq 1$,
   (d) $f(z) = z^3(z^2 + 1)^{-2}$, $\quad z \neq \pm i$.

1.4 SUFFICIENT CONDITIONS FOR ANALYTICITY

7. Using the Cauchy–Riemann equations, prove the following functions are nowhere differentiable:
   (a) $f(z) = \bar{z}$,
   (b) $f(z) = \operatorname{Re} z$,
   (c) $f(z) = \operatorname{Im} z$,
   (d) $f(z) = |z|$.
8. If $u(x, y) = e^x \cos y$, find a function $v(x, y)$ such that u and v satisfy the Cauchy–Riemann equations. (*Hint*: Apply the Cauchy–Riemann equation to u_x or u_y in order to find v.)
9. Using the chain rule prove that an entire function of an entire function is entire.
10. Prove $f(z) = z \operatorname{Im} z$ is differentiable only at $z = 0$, and find $f'(0)$.
11. Prove $f(z) = xy/|z|^2$ is continuous for $z \neq 0$. Can f be defined so as to make it continuous at $z = 0$?

1.4 SUFFICIENT CONDITIONS FOR ANALYTICITY

At this point one might ask whether the Cauchy–Riemann equations are enough to guarantee differentiability at a given point. The following example by Menchoff shows this is not the case. Let

$$f(z) = \begin{cases} \dfrac{z^5}{|z|^4}, & z \neq 0, \\ 0, & z = 0. \end{cases}$$

Then

$$\frac{f(z)}{z} = \left(\frac{z}{|z|}\right)^4, \quad z \neq 0,$$

which has value 1 on the real axis and value -1 on the line $y = x$. Thus f does not have a derivative at $z = 0$; but expanding the equation yields

$$u(x, 0) = x, \quad u(0, y) = 0 = v(x, 0), \quad v(0, y) = y,$$

hence

$$u_x(0, 0) = 1 = v_y(0, 0), \quad -u_y(0, 0) = 0 = v_x(0, 0),$$

and the Cauchy–Riemann equations hold.
However we do have the following:

Theorem Let $f(z) = u(x, y) + iv(x, y)$, defined in some domain G containing the point z_0, have continuous first partial derivatives with respect to x and y, satisfying the Cauchy–Riemann equations at z_0. Then $f'(z_0)$ exists.

Proof The difference quotient may be written in the form

$$\frac{f(z) - f(z_0)}{z - z_0} = \frac{u(x, y) - u(x_0, y_0)}{z - z_0} + i\frac{v(x, y) - v(x_0, y_0)}{z - z_0}$$

$$= \frac{x - x_0}{z - z_0}\left[\frac{u(x, y) - u(x_0, y)}{x - x_0} + i\frac{v(x, y) - v(x_0, y)}{x - x_0}\right]$$

$$+ \frac{y - y_0}{z - z_0}\left[\frac{u(x_0, y) - u(x_0, y_0)}{y - y_0} + i\frac{v(x_0, y) - v(x_0, y_0)}{y - y_0}\right]$$

$$= \frac{x - x_0}{z - z_0}[u_x(x_0 + t_1(x - x_0), y) + iv_x(x_0 + t_2(x - x_0), y)]$$

$$+ \frac{y - y_0}{z - z_0}[u_y(x_0, y_0 + t_3(y - y_0)) + iv_y(x_0, y_0 + t_4(y - y_0))]$$

where $0 < t_k < 1$, $k = 1, 2, 3, 4$, by the Mean-Value Theorem of differential calculus. Since the partials are continuous at z_0, we may write

$$\frac{f(z) - f(z_0)}{z - z_0} = \frac{x - x_0}{z - z_0}[u_x(z_0) + iv_x(z_0) + \varepsilon_1] + \frac{y - y_0}{z - z_0}[u_y(z_0) + iv_y(z_0) + \varepsilon_2],$$

where $\varepsilon_1, \varepsilon_2 \to 0$ as $z \to z_0$. Applying the Cauchy–Riemann equations to the last term we may combine the terms obtaining

$$\frac{f(z) - f(z_0)}{z - z_0} = u_x(z_0) + iv_x(z_0) + \frac{(x - x_0)\varepsilon_1 + (y - y_0)\varepsilon_2}{z - z_0}.$$

However,

$$\left|\frac{(x - x_0)\varepsilon_1 + (y - y_0)\varepsilon_2}{z - z_0}\right| \leq |\varepsilon_1| + |\varepsilon_2| \to 0 \quad \text{as} \quad z \to z_0,$$

hence the last term tends to 0 as z tends to z_0, so taking the limit we have

$$f'(z_0) = \lim_{z \to z_0} \frac{f(z) - f(z_0)}{z - z_0} = u_x(z_0) + iv_x(z_0).$$

In particular, if the hypotheses in the theorem hold at all points of the domain G, then f is analytic in G.

1.4 SUFFICIENT CONDITIONS FOR ANALYTICITY

Recall that for real functions $f'(x) \equiv 0$ implies $f(x)$ is constant. We prove the complex analog:

Theorem Let f be analytic on a domain G and $f'(z) = 0$ at each z in G. Then f is constant on G. The same conclusion holds if either $\operatorname{Re} f$, $\operatorname{Im} f$, $|f|$, or $\arg f$ is constant in G.

Proof Since $f'(z) = u_x(z) + iv_x(z)$, the vanishing of the derivative implies $u_x = v_y$, $v_x = -u_y$ are all zero. Thus u and v are constant on lines parallel to the coordinate axes, and since G is polygonally connected, $f = u + iv$ is constant on G.

If u (or v) is constant, $v_x = -u_y = 0 = u_x \; (=v_y)$ implying $f'(z) = u_x(z) + iv_x(z) = 0$ and f is constant.

If $|f|$ is constant, so is $u^2 + v^2$ implying that

$$uu_x + vv_x = 0, \qquad uu_y + vv_y = vu_x - uv_x = 0.$$

Solving for u_x, v_x, we have $u_x = v_x = 0$ unless the determinant $u^2 + v^2$ vanishes. Since $|f|^2 = u^2 + v^2$ is constant, if $u^2 + v^2 = 0$ at a single point, then it is constantly zero and f vanishes identically. Otherwise the derivative vanishes and f is constant.

If $\arg f = c$, then $f(G)$ lies on the line

$$v = (\tan c) \cdot u,$$

unless $u \equiv 0$ in which case we are done. But $(1 - i \tan c)f$ is analytic and

$$\operatorname{Im}(1 - i \tan c)f = v - (\tan c)u = 0,$$

implying $(1 - i \tan c)f$ is constant. Thus, so is f.

EXERCISES

1. Prove that each of these functions is entire:
   (a) $f(z) = e^x(\cos y + i \sin y)$,
   (b) $f(z) = \cos x \cosh y - i \sin x \sinh y$,
   (c) $f(z) = \sin x \cosh y + i \cos x \sinh y$.

2. Do analytic functions $f(z) = u + iv$ exist for which
   (a) $u = \dfrac{x}{x^2 + y^2}$,
   (b) $u = \log(x^2 + y^2)$,
   (c) $u = e^{y/x}$?
   If so indicate the domain of definition.

3. Show that at $z = 0$ the function

$$f(z) = \begin{cases} \dfrac{\bar{z}^3}{|z|^2}, & z \neq 0, \\ 0, & z = 0, \end{cases}$$

satisfies the Cauchy–Riemann equation but is not differentiable.

4. If u and v are expressed in terms of polar coordinates (r, θ), show that the Cauchy–Riemann equations can be written in the form

$$\frac{\partial u}{\partial r} = \frac{1}{r}\frac{\partial v}{\partial \theta}, \qquad \frac{1}{r}\frac{\partial u}{\partial \theta} = -\frac{\partial v}{\partial r}, \qquad r \neq 0.$$

5. If $f(z) = u + iv$ and $\bar{f} = u - iv$ are both analytic, prove f is constant.
6. Let $f(z) = u + iv$ be entire and suppose $u \cdot v$ is constant. Prove f is constant.
7. If $f(z) = u + iv$ is entire and $v = u^2$, then show that f is constant.

1.5 SOME ELEMENTARY FUNCTIONS

We have seen in Section 1.3 that polynomials and rational functions in a real variable yield analytic functions when the real variable is replaced by z. This is by no means an isolated example. In fact all elementary functions in calculus, such as exponentials, logarithms, trigonometric functions, give rise to analytic functions when suitably extended to the complex plane. In these next two sections we shall define extensions of these elementary functions and indicate some of their properties.

We begin with the exponential e^x. We wish to define a function $f(z) = e^z$ which is analytic and coincides with the real exponential function when z is real. Recalling that the real exponential is determined by the differential equation

$$f'(x) = f(x), \qquad f(0) = 1,$$

we ask if there is an analytic solution of the equation

$$f'(z) = f(z), \qquad f(0) = 1.$$

If such a solution exists, it will necessarily coincide with e^x when $z = x$ as it will satisfy the determining equation on the real axis. By the definition of f' we have

$$u_x + iv_x = u + iv, \qquad u(0) = 1, \qquad y(0) = 0.$$

1.5 SOME ELEMENTARY FUNCTIONS

Since $u_x = u$, $v_x = v$, integrating with respect to x we have

$$u(x, y) = p(y)e^x, \qquad v(x, y) = q(y)e^x,$$

with $p(0) = 1$, $q(0) = 0$ by the initial conditions. Differentiating these two equations with respect to y and applying the Cauchy–Riemann equations we obtain

$$p'(y)e^x = u_y = -v_x = -q(y)e^x, \qquad q'(y)e^x = v_y = u_x = p(y)e^x,$$

hence $p' = -q$, $q' = p$, so

$$q'' = p' = -q, \qquad p'' = -q' = -p,$$

and p, q are both solutions of the real differential equation $\phi''(y) + \phi(y) = 0$. All solutions of this equation are of the form $A \cos y + B \sin y$, with A and B constants. Since $q'(0) = p(0) = 1$, $p'(0) = -q(0) = 0$ we must have $p(y) = \cos y$, $q(y) = \sin y$. Hence

$$f(z) = e^x \cos y + i e^x \sin y = e^x(\cos y + i \sin y)$$

which coincides with e^x when $z = x$, and is analytic since the construction automatically guarantees the partials satisfy the Cauchy–Riemann equations and are continuous.

Definition The *complex exponential* given by

$$e^z = e^x(\cos y + i \sin y)$$

is a nonzero entire function satisfying the differential equation

$$f'(z) = f(z), \qquad f(0) = 1.$$

That $e^z \neq 0$ follows since neither e^x nor $\cos y + i \sin y$ vanishes. Observe further that

$$e^{iy} = \cos y + i \sin y, \qquad |e^{iy}| = 1.$$

Thus the polar representation of a complex number becomes

$$z = |z| e^{i \arg z}.$$

To visualize the mapping $w = e^z$ observe that the infinite strip $-\pi \leq y < \pi$ is mapped onto $\mathscr{C} - \{0\}$; the points on the line segment $x = 0$, $-\pi \leq y < \pi$

are mapped one-to-one onto the circle $|w| = 1$, the left half of the strip is mapped onto $0 < |w| < 1$, and the right half goes onto $|w| > 1$ (see Figure 1.11). Observe that e^z has period $2\pi i$

$$e^{z+2\pi i} = e^{x+(2\pi+y)i} = e^x[\cos(2\pi + y) + i \sin(2\pi + y)] = e^z,$$

so the complex values e^z and $e^{z+2\pi ik}$, k an integer, are identical. Hence, each infinite strip $-\pi \leq y - 2\pi k < \pi$, $k = 0, \pm 1, \pm 2, \ldots$, is also mapped onto $\mathscr{C} - \{0\}$, and the mapping $e^z \colon \mathscr{C} \to \mathscr{C} - \{0\}$ sends infinitely many points in $\mathscr{C}$ to the same point in $\mathscr{C} - \{0\}$. This is an undesirable development, since it

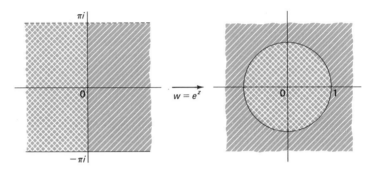

Figure 1.11
The exponential function.

prevents the discussion of an inverse function except on the infinite strips described above, and the inverse function is certain to be important because the inverse of the real exponential is the logarithm. To eliminate these difficulties, imagine the range of the mapping to consist of infinitely many copies of $\mathscr{C} - \{0\}$ stacked as layers one upon another, each cut along the negative real axis with the upper edge of one layer "glued" to the lower edge of the layer above (see Figure 1.12). Thus, we have a set $\mathscr{R}$ resembling an infinite spiraling ramp, and the mapping $e^z \colon \mathscr{C} \to \mathscr{R}$ sends each infinite strip continuously onto a layer of $\mathscr{R}$ (see Figure 1.12). The analytic mapping $e^z \colon \mathscr{C} \to \mathscr{R}$ is now one-to-one, and has an inverse which will be studied in the next section. The set $\mathscr{R}$ is called a *Riemann surface*; the cut lines on each copy of $\mathscr{C} - \{0\}$ are called *branch cuts*, the ends of the branch cuts $0, \infty$ are called *branch points*, and each copy of $\mathscr{C} - \{0\}$ is called a *branch* of $\mathscr{R}$. These notions will be made precise in Section 3.6 and are presented here primarily to develop the insight needed in later sections.

1.6 CONTINUATION

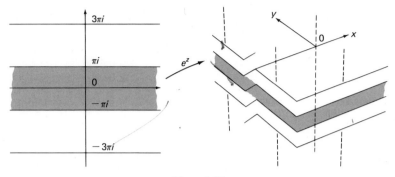

Figure 1.12
The Riemann surface of $w = e^z$.

EXERCISES

1. Show that:
   (a) $e^{z_1+z_2} = e^{z_1}e^{z_2}$,
   (b) $e^{z_1-z_2} = \dfrac{e^{z_1}}{e^{z_2}}$,
   (c) $e^{\bar{z}} = \overline{(e^z)}$,
   (d) $e^{nz} = (e^z)^n$, n an integer.

2. Obtain all values of $e^{\pi i k/2}$, k an integer.

3. Prove De Moivre's Theorem
$$(\cos\theta + i\sin\theta)^n = \cos n\theta + i\sin n\theta, \quad n \text{ an integer.}$$

4. Using De Moivre's Theorem, find the sums:
   (a) $1 + \cos x + \cos 2x + \cdots + \cos nx$,
   (b) $\cos x + \cos 3x + \cos 5x + \cdots + \cos(2n-1)x$,
   (c) $\sin x + \sin 2x + \sin 3x + \cdots + \sin nx$,
   (d) $\sin x + \sin 3x + \sin 5x + \cdots + \sin(2n-1)x$.

5. If $f(z)$ is entire, show $e^{f(z)}$ is entire and find its derivative.

6. Prove that e^z is the only analytic solution to the complex differential equation $f'(z) = f(z)$, $f(0) = 1$.

7. Find an analytic function mapping $\{z : 0 < x < 1, 0 \le y < 1\}$ one-to-one and onto $1 < |w| < e^{2\pi}$.

1.6 CONTINUATION

Since $e^z : \mathscr{C} \to \mathscr{R}$ is one-to-one, it has an inverse function mapping $\mathscr{R}$ onto $\mathscr{C}$. Imitating the real case we call this inverse mapping the *logarithm*. By definition it satisfies
$$\log(e^z) = z,$$

hence for any z in $\mathscr{R}$, we have
$$\log(z) = \log(|z|e^{i\arg z}) = \log(e^{\log|z|+i\arg z}) = \log|z| + i\arg z,$$
where arg z is now a single-valued real function on $\mathscr{R}$.

Theorem The function $\log z = \log|z| + i\arg z$ is analytic for all z in $\mathscr{R}$.

Proof Since $u = \log|z| = \tfrac{1}{2}\log(x^2 + y^2)$, $v = \arg z = \tan^{-1} y/x$,
$$u_x = \frac{x}{x^2+y^2}, \quad u_y = \frac{y}{x^2+y^2}, \quad v_x = \frac{-y}{x^2+y^2}, \quad v_y = \frac{x}{x^2+y^2},$$
so the Cauchy–Riemann equations hold and the partial derivatives are all continuous in $\mathscr{R}$. By the theorem in Section 1.4, $\log z$ is analytic in $\mathscr{R}$.

The logarithm possesses the following properties on $\mathscr{R}$:

(i) $e^{\log(z)} = e^{\log|z|+i\arg z} = |z|e^{i\arg z} = z.$

Applying the chain rule to (i) we have
$$1 = e^{\log z} \cdot (\log z)',$$
hence

(ii) $(\log z)' = 1/z.$

Visualizing the logarithm as the inverse mapping of the exponential, call the branch of $\mathscr{R}$ cut along the negative real axis, which is mapped onto the infinite strip $-\pi \le y < \pi$, the *principal branch* of the logarithm (see Figure 1.12). Denote $\log z$ when restricted to the principal branch by
$$\text{Log } z = \log|z| + i\,\text{Arg } z,$$
and call this the *principal value* of $\log z$.

The complex logarithm and exponential functions may be used to define the power functions:

Definition
$$z^a = e^{a\log z}, \quad a \text{ complex}, \quad z \neq 0.$$

The function $z^a: \mathscr{R} \to \mathscr{R}$ is analytic and one-to-one, as it is the composition of such functions. By the chain rule
$$(z^a)' = e^{a\log z} \cdot \frac{a}{z} = az^{a-1}.$$

1.6 CONTINUATION

The *principal value* of the power function is given by

$$z^a = e^{a \operatorname{Log} z}.$$

We are often interested in the case where $a = m/n > 0$, m, n positive integers with no common factors. Consider the set of numbers $e^{\operatorname{Log}(z) + 2\pi k i}$, $k = 0, \pm 1, \pm 2, \ldots$, that is, those points in $\mathscr{R}$ lying directly "above" and "below" the point $e^{\operatorname{Log} z}$. Then $(e^{\operatorname{Log}(z) + 2\pi k i})^{m/n} = e^{(m/n)\operatorname{Log} z} e^{(m/n) 2\pi k i}$ and writing $k = pn + q$ with p and q integers, $0 \le q < n$, we have

$$e^{(m/n) 2\pi k i} = e^{2\pi p m i} e^{2\pi i q m/n} = e^{2\pi i q m/n},$$

so there are only n different complex-valued answers. Thus, the mapping $z^{m/n} \colon \mathscr{R} \to \mathscr{C} - \{0\}$ repeats itself after n copies of $\mathscr{C} - \{0\}$. This fact makes it possible to simplify the model used in describing the mapping $w = z^{m/n}$. For simplicity suppose $m = 1$. Then

$$w = z^{1/n} = e^{(1/n) \operatorname{Log} z} e^{2\pi i q/n}, \qquad q = 0, 1, \ldots, n-1,$$

may be visualized as a mapping of $[\mathscr{C} - \{0\}]^n$ onto $[\mathscr{C} - [0]]$, where $[\mathscr{C} - \{0\}]^n$ consists of n copies of $\mathscr{C} - \{0\}$ "glued" one after another along the negative real axis, as in $\mathscr{R}$, except that the upper edge of the top branch is "glued" to the lower edge of the bottom branch (see Figure 1.13). If $\theta = \operatorname{Arg} z$, the point z on the principal branch is mapped onto the point $|z|^{1/n} e^{i\theta/n}$, and the points located where z is on the succeeding branches are mapped onto the circle $w = |z|^{1/n}$ at intervals of $2\pi/n$ radians after this point (see Figure 1.14). The mapping $z^n \colon [\mathscr{C} - \{0\}] \to [\mathscr{C} - \{0\}]^n$ is the inverse mapping and hence

$$z^{m/n} \colon [\mathscr{C} - \{0\}]^n \to [\mathscr{C} - \{0\}]^n$$

is analytic and one-to-one on the modified Riemann surfaces described above.

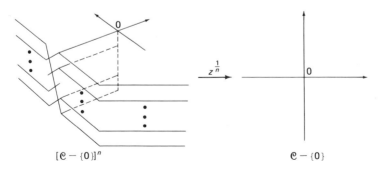

Figure 1.13
$z^{1/n} \colon [\mathscr{C} - \{0\}]^n \to \mathscr{C} - \{0\}$.

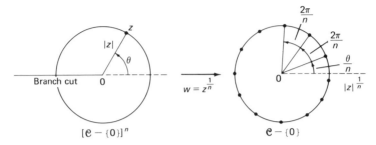

Figure 1.14

Since $e^{iy} = \cos y + i \sin y$ and $e^{-iy} = \cos y - i \sin y$ it follows that

$$\cos y = \frac{e^{iy} + e^{-iy}}{2}, \quad \sin y = \frac{e^{iy} - e^{-iy}}{2i}.$$

We extend these definitions to the complex plane:

Definition

$$\cos z = \frac{e^{iz} + e^{-iz}}{2}, \quad \sin z = \frac{e^{iz} - e^{-iz}}{2i}.$$

These functions are entire as they are sums of entire functions and satisfy

$$(\cos z)' = \frac{ie^{iz} - ie^{-iz}}{2} = -\frac{e^{iz} - e^{-iz}}{2i} = -\sin z,$$

$$(\sin z)' = \frac{ie^{iz} + ie^{-iz}}{2i} = \frac{e^{iz} + e^{-iz}}{2} = \cos z.$$

The other four trigonometric functions, defined in terms of the sine and cosine function by the usual relations

$$\tan z = \frac{\sin z}{\cos z}, \quad \cot z = \frac{\cos z}{\sin z},$$

$$\sec z = \frac{1}{\cos z}, \quad \csc z = \frac{1}{\sin z},$$

are analytic except where their denominators vanish, and satisfy the usual rules of differentiation

$$(\tan z)' = \sec^2 z, \quad (\sec z)' = \sec z \tan z,$$

$$(\cot z)' = -\csc^2 z, \quad (\csc z)' = -\csc z \cot z.$$

1.6 CONTINUATION

All the usual identities are still valid in complex variables, the proofs depending on properties of the exponential, for example

$$\cos^2 z + \sin^2 z = \tfrac{1}{4}[(e^{iz} + e^{-iz})^2 - (e^{iz} - e^{-iz})^2] = 1.$$

From the definition of $\cos z$ we have

$$\cos z = \cos(x + iy)$$

$$= \frac{e^{-y+ix} + e^{y-ix}}{2}$$

$$= \frac{1}{2} e^{-y}(\cos x + i \sin x) + \frac{1}{2} e^{y}(\cos x - i \sin x)$$

$$= \left(\frac{e^y + e^{-y}}{2}\right) \cos x - i\left(\frac{e^y - e^{-y}}{2}\right) \sin x.$$

Thus

$$\cos z = \cos x \cosh y - i \sin x \sinh y.$$

Similarly we find

$$\sin z = \sin x \cosh y + i \cos x \sinh y.$$

Theorem The real zeros of $\sin z$ and $\cos z$ are their only zeros.

Proof If $\sin z = 0$, the last equation shows we must have

$$\sin x \cosh y = 0, \qquad \cos x \sinh y = 0.$$

But $\cosh y \geq 1$ implying the first term vanishes only when $\sin x = 0$, that is $x = 0, \pm\pi, \pm 2\pi, \ldots$. However, for these values $\cos x$ does not vanish. Hence we must have $\sinh y = 0$, or $y = 0$. Thus

$$\sin z = 0 \quad \text{implies} \quad z = n\pi, \qquad n \text{ an integer.}$$

This statement also applies to $\tan z$, and in like manner we find

$$\cos z = 0 \quad \text{implies} \quad z = (n + \tfrac{1}{2})\pi, \qquad n \text{ an integer.}$$

The hyperbolic functions are defined in an analogous fashion

$$\sinh z = \frac{e^z - e^{-z}}{2}, \qquad \cosh z = \frac{e^z + e^{-z}}{2}.$$

The usual identities and rules for differentiation apply here too. In fact most of the mathematical functions arising in physical and engineering problems are analytic. Thus the concept of differentiation applies to a large useful class of functions.

EXERCISES

1. Show that:

   (a) $\log z_1 + \log z_2 = \log z_1 z_2$,

   (b) $\log z_1 - \log z_2 = \log \dfrac{z_1}{z_2}$,

   (c) $z^a z^b = z^{a+b}$,

   (d) $\dfrac{z^a}{z^b} = z^{a-b}$.

2. Find the principal values of

   (a) $\log i$,   (b) $\log(1 + i)$,
   (c) i^i,   (d) $(1 + i)^{1+i}$.

3. For what complex numbers a can z^a be extended continuously at $z = 0$? When is the resulting function entire?

4. Prove that $\log z$ is the only analytic solution of the differential equation

$$f'(z) = \frac{1}{z}, \qquad f(1) = 0,$$

   in the disk $|z - 1| < 1$.

5. Prove the identities:

   (a) $\sin(z_1 + z_2) = \sin z_1 \cos z_2 + \cos z_1 \sin z_2$,
   (b) $\cos(z_1 + z_2) = \cos z_1 \cos z_2 - \sin z_1 \sin z_2$,
   (c) $\sin(-z) = -\sin z$, $\cos(-z) = \cos z$,
   (d) $\sin 2z = 2 \sin z \cos z$, $\cos 2z = \cos^2 z - \sin^2 z$,

$$\tan 2z = \frac{2 \tan z}{1 - \tan^2 z},$$

   (e) $|\sin z|^2 = \sin^2 x + \sinh^2 y$,
   (f) $|\cos z|^2 = \cos^2 x + \sinh^2 y$.

6. Prove the rules of differentiation for the functions $\tan z$, $\cot z$, $\sec z$, $\csc z$ are valid as stated.

NOTES

7. Prove the identities:
  (a) $\cosh^2 z - \sinh^2 z = 1$,
  (b) $\sinh(z_1 + z_2) = \sinh z_1 \cosh z_2 + \cosh z_1 \sinh z_2$,
  (c) $\cosh(z_1 + z_2) = \cosh z_1 \cosh z_2 + \sinh z_1 \sinh z_2$,
  (d) $i \sinh z = \sin(iz)$, $\cosh z = \cos(iz)$, $i \tanh z = \tan(iz)$,
  (e) $|\sinh z|^2 = \sinh^2 x + \sin^2 y$, $|\cosh z|^2 = \sinh^2 x + \cos^2 y$.

8. Prove the following rules of differentiation:
  (a) $(\sinh z)' = \cosh z$, $(\cosh z)' = \sinh z$,
  (b) $(\tanh z)' = \operatorname{sech}^2 z$, $(\coth z)' = -\operatorname{csch}^2 z$,
  (c) $(\operatorname{sech} z)' = -\operatorname{sech} z \tanh z$, $(\operatorname{csch} z)' = -\operatorname{csch} z \coth z$.

9. Find all the zeros of $\sinh z$ and $\cosh z$.

10. Show that the function $w = \sin z$ maps
  (a) the strip $|x| < \pi/2$ onto $\mathscr{C} - \{z: y = 0, |x| \geq 1\}$,
  (b) the semiinfinite strip $|x| < \pi/2$, $y > 0$ onto the upper half plane,
  (c) the semiinfinite strip $0 < x < \pi/2$, $y > 0$ onto the first quadrant,

by indicating what happens to horizontal and vertical line segments under the transformation

$$w = \sin z = \sin x \cosh y + i \cos x \sinh y.$$

NOTES

Section 1.1 Formulas relating z to Z in the stereographic projection are easy to compute: [A, pp. 18–20] or [H, pp. 38–44].

Section 1.3 Other synonyms for *analytic* are *holomorphic*, *monogenic*, and *regular*.

Section 1.4 Far weaker sufficient conditions are known for analyticity. The best such result appears to be in [S, pp. 197–199], where the *Looman–Menchoff* theorem states that if u and v are continuous in G, have first partials at all except an enumerable number of points in G, and satisfy the Cauchy–Riemann equations almost everywhere in G, then $f = u + iv$ is analytic in G. Other examples showing the Cauchy–Riemann equations alone are insufficient for analyticity may be found in [T, pp. 67, 70].

Sections 1.5 and 1.6 For a more detailed elementary development of Riemann surfaces see [Kn, Part II, pp. 100–146]. Tables of elementary mappings of domains can be found in the Appendix and in [Ko].

Chapter 2 | COMPLEX INTEGRATION

2.1. LINE INTEGRALS

The properties of analytic functions discussed in the preceding chapter were all consequences of the differentiability of the function. In real calculus, the Fundamental Theorem of Integration reveals a surprising and useful connection between derivatives and definite integrals. One of our main goals will be to prove the same theorem for line integrals in the complex plane. At first glance this appears to be a very difficult job as there is an infinity of curves joining two given points, but the proof is easy and the applications are very useful.

2.1 LINE INTEGRALS

The equation of an *arc* γ in the plane is most conveniently given in parametric form

$$\gamma: x = x(t), \qquad y = y(t), \qquad \alpha \le t \le \beta,$$

with $x(t)$, $y(t)$ continuous functions of the real variable t in the closed real interval $[\alpha, \beta]$. In the complex plane we use the notation

$$\gamma: z = z(t) = x(t) + iy(t), \qquad \alpha \le t \le \beta.$$

We say γ is *differentiable* if $z'(t) = x'(t) + iy'(t)$ exists and is continuous, and γ is said to be *piecewise differentiable* (pwd) if it is differentiable except at a finite number of values t at which $z(t)$ is continuous and has left and right derivatives. An arc is *simple*, or a *Jordan arc*, if $z(t_1) = z(t_2)$ only if $t_1 = t_2$, that is, it is nonself-intersecting. An arc is a *closed curve* if $z(\alpha) = z(\beta)$, and is a *Jordan curve* if it is closed and simple except at the *endpoints* α and β. Jordan curves satisfy the following property:

Jordan Curve Theorem A Jordan curve separates the extended plane into two simply-connected domains, both having the curve as their boundary.

The domain containing the point at infinity is called the *outside* of the curve, the other domain is called the *inside*. Although this theorem seems obvious, a formal proof is long and tedious, so we shall accept its validity on intuitive grounds. A Jordan curve is parametrized in its *positive sense* if its interior is kept to the left as the curve is traversed. For example, the parametrization $z(t) = e^{it} = \cos t + i \sin t$, $0 \le t \le 2\pi$ parametrizes $|z| = 1$ in its positive sense, whereas $z(t) = e^{-it}$, $0 \le t \le 2\pi$ does not.

An *opposite arc* of $\gamma: z = z(t)$, $\alpha \le t \le \beta$, is the arc

$$-\gamma: z = z(-t), \qquad -\beta \le t \le -\alpha.$$

We shall define our line integrals by making use of the parametrization and thus reducing them to integrals on the real line. This allows us to employ the integration theory developed in ordinary calculus.

Definition Let $\gamma: z = z(t)$, $\alpha \le t \le \beta$, be a differentiable arc and $f(z) = u + iv$ be continuous on γ. Then

$$\int_\gamma f(z)\,dz = \int_\alpha^\beta f(z(t))z'(t)\,dt$$

$$= \int_\alpha^\beta [u(z(t)) + iv(z(t))][x'(t) + iy'(t)]\,dt$$

$$= \int_\alpha^\beta [u(z(t))x'(t) - v(z(t))y'(t)]\,dt$$

$$+ i\int_\alpha^\beta [u(z(t))y'(t) + v(z(t))x'(t)]\,dt.$$

The line integral over a pwd arc γ is obtained by applying the definition above to each of the finitely many closed intervals on which $z(t)$ is differentiable, and summing the results.

It is important to note that the value of the integral above is independent of the parametrization of γ. Any change of parameters is determined by a piecewise differentiable increasing function $t = t(T)$ mapping $[A, B]$ onto $[\alpha, \beta]$. Changing variables, and using the chain rule, we have

$$\int_\alpha^\beta f(z(t))z'(t)\,dt = \int_A^B f(z(t(T)))z'(t(T))t'(T)\,dT = \int_A^B f(z(T))z'(T)\,dT.$$

Thus it is immaterial which parametrization is employed.

Integrals have the following useful properties:

(i) $\int_\gamma [\alpha f_1(z) + \beta f_2(z)]\,dz = \alpha \int_\gamma f_1(z)\,dz + \beta \int_\gamma f_2(z)\,dz$.
(ii) $\int_{\gamma_1 + \gamma_2} f(z)\,dz = \int_{\gamma_1} f(z)\,dz + \int_{\gamma_2} f(z)\,dz$, where $\gamma_1 + \gamma_2$ is the path consisting of traversing first γ_1 followed by γ_2.
(iii) $\int_{-\gamma} f(z)\,dz = -\int_\gamma f(z)\,dz$.
(iv) $|\int_\gamma f(z)\,dz| \leq \int_\gamma |f(z)|\,|dz|$,

where we define $|dz|$ to be the differential with respect to *arc length*, with

$$|dz| = |dx + idy| = \sqrt{(dx)^2 + (dy)^2} = ds.$$

To prove (iv) notice that

$$\text{Re}\left(e^{-i\theta}\int_\gamma f(z)\,dz\right) = \int_\alpha^\beta \text{Re}(e^{-i\theta}f(z(t))z'(t))\,dt \leq \int_\alpha^\beta |f(z(t))|\,|z'(t)|\,dt.$$

For $\theta = \arg(\int_\gamma f(z)\,dz)$ the expression on the left reduces to the absolute value of the integral and the inequality holds. The remaining proofs are left as exercises.

2.1 LINE INTEGRALS

The first two examples are important and will be used later:

Example 1

$$\int_{|z|=1} \frac{dz}{z} = 2\pi i.$$

Unless the contrary is stated, integrations over Jordan curves are assumed to be carried out in the positive sense. Thus parametrizing $|z| = 1$ by $z(t) = e^{it}$, $0 \le t \le 2\pi$, we have

$$dz = z'(t)\, dt = ie^{it}\, dt.$$

Then our integral becomes

$$\int_0^{2\pi} \frac{ie^{it}}{e^{it}}\, dt = i\int_0^{2\pi} dt = 2\pi i.$$

Example 2 Let γ be a differentiable arc joining z_1 to z_2 in $\mathscr{C}$. Then

$$\int_\gamma k\, dz = k(z_2 - z_1), \qquad \int_\gamma z\, dz = \tfrac{1}{2}(z_2^2 - z_1^2), \qquad k \text{ a constant.}$$

Parametrizing γ by $z = z(t)$, $\alpha \le t \le \beta$, we obtain $dz = z'(t)\, dt$ and

$$\int_\gamma k\, dz = k\int_\alpha^\beta z'(t)\, dt = kz(t)\Big|_\alpha^\beta = k(z_2 - z_1),$$

$$\int_\gamma z\, dz = \int_\alpha^\beta z(t) z'(t)\, dt = \tfrac{1}{2}z^2(t)\Big|_\alpha^\beta = \tfrac{1}{2}(z_2^2 - z_1^2).$$

Note that the conclusions are valid for any pwd arc joining z_1 to z_2 and that both integrals vanish if γ is a pwd closed curve. Both are proved by partitioning γ into a finite number of differentiable arcs on each of which the integrals are evaluated, and then summing up the answers.

Example 3 To evaluate the integral $\int |z|\, dz$ along the straight line γ joining 0 to $1 + i$, parametrize γ by

$$\gamma : z(t) = t + it, \qquad 0 \le t \le 1.$$

Then $z'(t) = 1 + i$, so by the definition

$$\int_\gamma |z|\, dz = \int_0^1 |t + it|(1 + i)\, dt = \sqrt{2}(1 + i)\int_0^1 t\, dt = \frac{1 + i}{\sqrt{2}}.$$

Example 4

$$\left| \int_{|z|=1} e^z \, dz \right| \leq 2\pi e.$$

From property (iv) we have

$$\left| \int_{|z|=1} e^z \, dz \right| \leq \int_{|z|=1} |e^z| \, |dz|.$$

Since $|e^z| = e^x \leq e$ for all points $z = x + iy$ on the unit circle,

$$\int_{|z|=1} |e^z| \, |dz| \leq e \int_{|z|=1} |dz| = 2\pi e,$$

and the inequality is verified. In fact, it is clear that

$$\left| \int_{|z|=1} e^z \, dz \right| < 2\pi e,$$

since $|e^z|$ attains the value e only at $z = 1$.

EXERCISES

1. Prove the integral identities:
   (a) $\int_\gamma [\alpha f_1(z) + \beta f_2(z)] \, dz = \alpha \int_\gamma f_1(z) \, dz + \beta \int_\gamma f_2(z) \, dz$,
   (b) $\int_{\gamma_1 + \gamma_2} f(z) \, dz = \int_{\gamma_1} f(z) \, dz + \int_{\gamma_2} f(z) \, dz$,
   (c) $\int_{-\gamma} f(z) \, dz = -\int_\gamma f(z) \, dz$.

2. Suppose $|f(z)| \leq M$ at every point z on the pwd arc γ of length L. Show that

$$\left| \int_\gamma f(z) \, dz \right| \leq ML.$$

3. Let $P(z)$ be any polynomial and γ be a pwd arc. Show that:
   (a) $\int_\gamma P(z) \, dz = 0$, if γ is a closed curve,
   (b) $\int_\gamma P(z) \, dz$ depends only on the endpoints of γ.

4. Evaluate the integrals

$$\int x \, dz, \quad \int y \, dz, \quad \int \bar{z} \, dz,$$

along the following paths:

   (a) the directed line segment from 0 to $1 - i$,
   (b) around the circle $|z| = 1$,
   (c) around the circle $|z - a| = R$.

2.2 THE CAUCHY–GOURSAT THEOREM

5. Evaluate the integral $\int (z-a)^n \, dz$, n an integer, around the circle $|z-a| = R$. (The answer for $n = -1$ differs from the rest.)
6. Without computing the integral, show that

$$\left| \int_{|z|=2} \frac{dz}{z^2 + 1} \right| \leq \frac{4\pi}{3}.$$

7. If γ is the semicircle $|z| = R$, $|\arg z| \leq \pi/2$, $R > 1$, show that

$$\left| \int_\gamma \frac{\operatorname{Log} z}{z^2} \, dz \right| \leq \frac{\pi}{R} \left(\operatorname{Log} R + \frac{\pi}{2} \right),$$

and hence that the value of the integral tends to zero as $R \to \infty$.
8. Evaluate $\int_\gamma e^z \, dz$, where γ is:

   (a) the straight line path joining 1 to i,
   (b) the path is the first quadrant along the circle $|z| = 1$ joining 1 to i,
   (c) the path along the coordinate axes joining 1 to i.

9. If γ is the ellipse $z(t) = a \cos t + ib \sin t$, $0 \leq t \leq 2\pi$, $a^2 - b^2 = 1$, show that

$$\int_\gamma \frac{dz}{\sqrt{1 - z^2}} = \pm 2\pi,$$

depending on which value of the radical is taken.
10. Evaluate $\int_{|z|=1} |z + 1| \, |dz|$.

2.2 THE CAUCHY–GOURSAT THEOREM

In Exercise 3(a) and Example 1 of the last section we found that the line integral of a polynomial along a pwd closed curve vanishes, but that

$$\int_{|z|=1} \frac{dz}{z} = 2\pi i.$$

However, the function $1/z$ is not analytic at the origin. Might it be true that the line integral of a function along a pwd Jordan curve vanishes when the function is analytic on and inside the curve? Surprisingly, the answer is in the affirmative. The first step in obtaining this result is the following theorem, due to Cauchy and Goursat:

Theorem Let the function $f(z)$ be analytic on a domain containing the rectangle R, given by the inequalities $a \leq x \leq b$, $c \leq y \leq d$. Then

$$\int_{\partial R} f(z)\, dz = 0,$$

where ∂R is the boundary of R.

Proof To simplify the notation let

$$I(R) = \int_{\partial R} f(z)\, dz$$

for any rectangle R. Bisect R into four congruent rectangles R^1, R^2, R^3, R^4, and observe that

$$I(R) = I(R^1) + I(R^2) + I(R^3) + I(R^4),$$

because the integrals over the common sides cancel each other, by property (iii), as they have opposite orientation (see Figure 2.1). By the Triangle

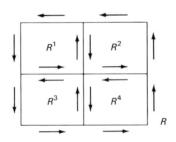

Figure 2.1

Inequality we have

$$|I(R)| \leq |I(R^1)| + |I(R^2)| + |I(R^3)| + |I(R^4)|,$$

so at least one R^j satisfies $|I(R^j)| \geq |I(R)|/4$. More than one R^j may have this property; select the one with smallest superscript and call it R_1. Repeating the above process indefinitely, we obtain a nested sequence of rectangles $R \supset R_1 \supset \cdots \supset R_n \supset R_{n+1} \supset \cdots$ satisfying

$$|I(R_n)| \geq \frac{|I(R_{n-1})|}{4},$$

implying that

$$|I(R_n)| \geq \frac{|I(R)|}{4^n}$$

2.2 THE CAUCHY–GOURSAT THEOREM

(see Figure 2.2). Denote by $z_n^* = x_n^* + iy_n^*$ the lower left corner of the rectangle R_n. From the construction of the rectangles R_n, it is clear that the sequences $\{x_n^*\}$ and $\{y_n^*\}$ of real numbers are nondecreasing and bounded above by b and d, respectively. Thus their limits x^* and y^* exist. We shall show that the point $z^* = x^* + iy^*$ belongs to all the rectangles R_n. If $z_n = x_n + iy_n$ is the upper right corner of R_n, then x_n and y_n are upper bounds for the sequences $\{x_n^*\}$ and $\{y_n^*\}$ implying that $x_n^* \le x^* \le x_n$, $y_n^* \le y^* \le y_n$. Thus z^* lies in R_n, for all n. Moreover, no other point lies in all the rectangles R_n, since $|z_n - z_n^*| \to 0$ as $n \to \infty$.

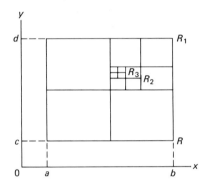

Figure 2.2

$R \supset R_1 \supset R_2 \supset R_3 \supset \cdots$.

Given $\varepsilon > 0$ we can find a $\delta > 0$ such that $f(z)$ is analytic and

$$\left| \frac{f(z) - f(z^*)}{z - z^*} - f'(z^*) \right| < \varepsilon,$$

whenever $|z - z^*| < \delta$. For sufficiently large n we have R_n contained in $|z - z^*| < \delta$. Since z^*, $f(z^*)$, $f'(z^*)$ are constants, Example 2 of the last section implies that

$$\int_{\partial R_n} f(z^*)\, dz = 0 = \int_{\partial R_n} f'(z^*)(z - z^*)\, dz,$$

thus, adding zero to the integral $I(R_n)$, we have

$$|I(R_n)| = \left| \int_{\partial R_n} [f(z) - f(z^*) - f'(z^*)(z - z^*)]\, dz \right|.$$

By property (iv) of integrals and the conditions above, we have

$$|I(R_n)| \leq \int_{\partial R_n} |f(z) - f(z^*) - f'(z^*)(z - z^*)| \, |dz|$$

$$< \varepsilon \int_{\partial R_n} |z - z^*| \, |dz| \leq \varepsilon D_n L_n,$$

where

$$D_n = |z_n - z_n^*|, \qquad L_n = \int_{\partial R_n} |dz|,$$

are the diagonal and length of the perimeter of R_n, respectively. But

$$D_n = \tfrac{1}{2} D_{n-1} = \cdots = 2^{-n} D, \qquad L_n = \tfrac{1}{2} L_{n-1} = \cdots = 2^{-n} L,$$

where D and L are the diagonal and length of the perimeter of R, so

$$4^{-n} |I(R)| \leq |I(R_n)| \leq \varepsilon D_n L_n = 4^{-n} \varepsilon DL.$$

Thus $|I(R)| \leq \varepsilon DL$, and since ε is arbitrary we can only have $I(R) = 0$, and the proof is complete.

The example and exercises below are nontrivial applications of the Cauchy–Goursat Theorem. The reader may prefer to read the next section before studying them.

Example †

$$\int_{-\infty}^{\infty} e^{-x^2} \cos 2bx \, dx = \sqrt{\pi} e^{-b^2}.$$

Applying the Cauchy–Goursat Theorem to the function $f(z) = e^{-z^2}$ analytic on a domain containing the rectangle $|x| \leq a$, $0 \leq y \leq b$ (see Figure 2.3), we have

$$0 = \int_{-a}^{a} e^{-x^2} \, dx + \int_{0}^{b} e^{-(a+iy)^2} i \, dy + \int_{a}^{-a} e^{-(x+ib)^2} \, dx + \int_{b}^{0} e^{-(-a+iy)^2} i \, dy$$

$$= \int_{-a}^{a} e^{-x^2} \, dx - e^{b^2} \int_{-a}^{a} e^{-x^2} (\cos 2bx - i \sin 2bx) \, dx$$

$$- i e^{-a^2} \int_{0}^{b} e^{y^2} (e^{2iay} - e^{-2iay}) \, dy$$

$$= \int_{-a}^{a} e^{-x^2} \, dx - e^{b^2} \int_{-a}^{a} e^{-x^2} \cos 2bx \, dx + 2 e^{-a^2} \int_{0}^{b} e^{y^2} \sin 2ay \, dy, \qquad (1)$$

2.2 THE CAUCHY–GOURSAT THEOREM

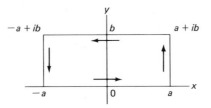

Figure 2.3

since the imaginary part of the middle integral vanishes. But, using polar coordinates,

$$\left(\int_{-\infty}^{\infty} e^{-x^2}\, dx\right)^2 = \int_{-\infty}^{\infty} e^{-x^2}\, dx \cdot \int_{-\infty}^{\infty} e^{-y^2}\, dy$$

$$= \int_{-\infty}^{\infty}\int_{-\infty}^{\infty} e^{-(x^2+y^2)}\, dx\, dy$$

$$= \int_0^{2\pi}\int_0^{\infty} e^{-r^2} r\, dr\, d\theta = \pi, \tag{2}$$

so the first two integrals in (1) are convergent as $a \to \infty$. Letting $a \to \infty$, the last integral in (1) vanishes and

$$\int_{-\infty}^{\infty} e^{-x^2} \cos 2bx\, dx = e^{-b^2} \int_{-\infty}^{\infty} e^{-x^2}\, dx = \sqrt{\pi}\, e^{-b^2}.$$

EXERCISES†

1. Assuming $0 < b < 1$ and applying the Cauchy–Goursat Theorem to the function $f(z) = (1 + z^2)^{-1}$ along the boundary of the same rectangle as in the example show that

$$\int_{-\infty}^{\infty} \frac{(1 - b^2 + x^2)\, dx}{(1 - b^2 + x^2)^2 + 4x^2 b^2} = \pi.$$

2. Prove that

$$\int_{-\infty}^{\infty} e^{-kx^2} \cos ax\, dx = \sqrt{\frac{\pi}{k}}\, e^{-a^2/4k}, \qquad k > 0, \quad a \text{ real},$$

by using the same procedure as in the example with the function $f(z) = e^{-kz^2}$. Check your answer by changing variables.

3. Prove that

$$\int_{-\infty}^{\infty} \frac{(1-b^2+x^2)\cos kx + 2xb \sin kx}{(1-b^2+x^2)^2 + 4x^2 b^2} dx = e^{kb} \int_{-\infty}^{\infty} \frac{\cos kx}{1+x^2} dx,$$

$$\int_{-\infty}^{\infty} \frac{(1-b^2+x^2)\sin kx - 2xb \cos kx}{(1-b^2+x^2)^2 + 4x^2 b^2} dx = 0,$$

with $0 < b < 1$ and k real.

4. Let $0 < b < 1$ and show that

$$\int_{-\infty}^{\infty} \frac{\operatorname{Re}(1+(x-ib)^4)}{|1+(x+ib)^4|^2} dx = \int_{-\infty}^{\infty} \frac{dx}{1+x^4}.$$

5. Prove that

$$\int_0^{\infty} e^{-x^2} \sin 2xb\, dx = e^{-b^2} \int_0^b e^{x^2} dx, \qquad b > 0,$$

by integrating around a suitable rectangle.

2.3 THE FUNDAMENTAL THEOREM OF INTEGRATION

As a second step in answering the question posed in the last section we show analytic functions have antiderivatives:

Theorem If $f(z)$ is analytic in the disk $|z - z_0| < r$, then there is a function $F(z)$ analytic in $|z - z_0| < r$ satisfying $F'(z) = f(z)$.

Proof For any z in the disk $|z - z_0| < r$ let γ_z be the arc consisting of the line segments joining z_0 to $x + iy_0$ and $x + iy_0$ to z, where $z = x + iy$, $z_0 = x_0 + iy_0$ (see Figure 2.4). Define

$$F(z) = \int_{\gamma_z} f(z)\, dz = \int_{x_0}^{x} f(t + iy_0)\, dt + i \int_{y_0}^{y} f(x + it)\, dt. \tag{1}$$

If γ_z' is the arc consisting of the line segments joining z_0 to $x_0 + iy$ and $x_0 + iy$ to z, then $\gamma_z - \gamma_z'$ is the boundary of a rectangle, and by the Cauchy–Goursat Theorem

$$0 = \int_{\gamma_z - \gamma_z'} f(z)\, dz = \int_{\gamma_z} f(z)\, dz - \int_{\gamma_z'} f(z)\, dz.$$

2.3 THE FUNDAMENTAL THEOREM OF INTEGRATION

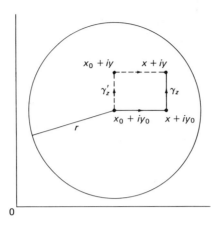

Figure 2.4

Thus we also can compute $F(z)$ along the path γ'_z,

$$F(z) = \int_{\gamma'_z} f(z)\, dz = i \int_{y_0}^{y} f(x_0 + it)\, dt + \int_{x_0}^{x} f(t + iy)\, dt. \qquad (2)$$

The partial derivative of (1) with respect to y is given by

$$F_y(z) = i \frac{\partial}{\partial y} \int_{y_0}^{y} f(x + it)\, dt = if(x + iy) = if(z),$$

since the first integral in (1) is independent of y. Similarly, taking the partial derivative of (2) with respect to x yields $F_x(z) = f(z)$. Thus $F(z)$ satisfies the Cauchy–Riemann equations

$$F_x(z) = f(z) = -iF_y(z),$$

and since $f(z)$ is continuous (see Exercise 4 of Section 1.3) we have sufficient conditions for the analyticity of $F(z)$ in $|z - z_0| < r$. Finally $F'(z) = F_x(z) = f(z)$.

As in real calculus, *two antiderivatives of the same function differ at most by a constant*: If $F(z)$ and $H(z)$ are both antiderivatives of the function $f(z)$ then

$$[F(z) - H(z)]' = f(z) - f(z) = 0,$$

implying $F(z) - H(z)$ is a constant, by the theorem on page 17.

Fundamental Theorem of Integration Let $f(z)$ be analytic and possess an analytic antiderivative $F(z)$ in a domain G. Then for any pwd arc $\gamma: z = z(t)$, $\alpha \le t \le \beta$,

$$\int_\gamma f(z)\, dz = F(z(\beta)) - F(z(\alpha)).$$

Proof Since $F(z)$ is analytic and is the antiderivative of $f(z)$, the Cauchy-Riemann equations imply $F_x = f = -iF_y$. Letting $f = u + iv$, $F = U + iV$, we have

$$U_x + iV_x = u + iv = V_y - iU_y,$$

so

$$\int_\gamma f(z)\, dz = \int_\gamma (u + iv)(dx + i\, dy)$$

$$= \int_\gamma u\, dx - v\, dy + i \int_\gamma v\, dx + u\, dy$$

$$= \int_\alpha^\beta [U_x \cdot x'(t) + U_y \cdot y'(t)]\, dt + i \int_\alpha^\beta [V_x \cdot x'(t) + V_y \cdot y'(t)]\, dt$$

$$= \int_{z(\alpha)}^{z(\beta)} dU + i \int_{z(\alpha)}^{z(\beta)} dV = F \Big|_{z(\alpha)}^{z(\beta)}.$$

Corollary Let $f(z)$ be analytic and γ be any pwd closed curve in the disk $|z - z_0| < r$. Then

$$\int_\gamma f(z)\, dz = 0.$$

Proof The corollary follows directly from the last two theorems, since $f(z)$ has an analytic antiderivative and the endpoints of γ coincide.

The next theorem is a generalization of the corollary to simply connected domains, and yields a complete answer to the question posed in Section 2.2:

Cauchy's Theorem If $f(z)$ is analytic and γ is a pwd closed curve in the simply connected domain G, then

$$\int_\gamma f(z)\, dz = 0.$$

2.3 THE FUNDAMENTAL THEOREM OF INTEGRATION

Proof By the Fundamental Theorem we need only show $f(z)$ has an analytic antiderivative in G. Fix a point z_0 in G, then by the theorem on page 9 we can find a polygon, with sides parallel to the axes, joining z_0 to any point z in G. Suppose γ and γ' are two such polygons, then $\gamma - \gamma'$ consists of a finite number of boundaries of rectangles in G (possibly some of them degenerate) traversed alternatively in the positive and negative direction (see Figure 2.5).

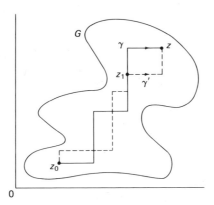

Figure 2.5

The curve $\gamma - \gamma'$ consists of the boundaries of rectangles in G.

This fact requires a delicate proof, utilizing the simple connectedness of the domain G, and is omitted as it is intuitively clear (see the Notes at the end of the chapter). By the Cauchy–Goursat theorem

$$0 = \int_{\gamma-\gamma'} f(z)\,dz = \int_{\gamma} f(z)\,dz - \int_{\gamma'} f(z)\,dz,$$

thus

$$F(z) = \int_{\gamma} f(z)\,dz = \int_{\gamma'} f(z)\,dz$$

is independent of the choice of path. Let the last line segment of γ (γ') be horizontal (vertical) and $z_1 = x_1 + iy_1$ be the last point of intersection. Then

$$F(z) = i\int_{y_1}^{y} f(x_1 + it)\,dt + \int_{x_1}^{x} f(t + iy)\,dt + C$$

$$= \int_{x_1}^{x} f(t + iy_1)\,dt + i\int_{y_1}^{y} f(x + it)\,dt + C,$$

where $z = x + iy$ and the constant $C = F(z_1)$. Taking the partial derivatives, the first equation yields $F_x(z) = f(z)$, the second $F_y(z) = if(z)$. Since $f(z)$ is continuous and $F_x = -iF_y$, $F(z)$ is analytic in G and $F'(z) = f(z)$.

In many applications it is necessary to consider domains which are not simply connected. We shall generalize Cauchy's Theorem to the case of these *multiply connected* domains:

Theorem Let the inside of the pwd Jordan curve γ_0 contain the disjoint pwd Jordan curves $\gamma_1, \ldots, \gamma_n$. Suppose $f(z)$ is analytic in a domain G containing the set S consisting of all points on and inside γ_0 but not inside γ_k, $k = 1, \ldots, n$. Then

$$\int_{\gamma_0} f(z)\, dz = \sum_{k=1}^{n} \int_{\gamma_k} f(z)\, dz.$$

Proof One can always find disjoint pwd arcs L_k, $k = 0, \ldots, n$, joining γ_k to γ_{k+1} (with L_n joining γ_n to γ_0) such that two pwd Jordan curves are formed each lying in some simply connected subdomain of G. (We omit a proof on intuitive grounds. See Figure 2.6.) By Cauchy's Theorem the integral of $f(z)$ on these curves, each traversed in the positive sense, vanishes. But the total contribution of these two curves is equivalent to traversing γ_0 in the positive sense, $\gamma_1, \ldots, \gamma_n$ in the negative (opposite) sense, and $L_0, \ldots, L_n$

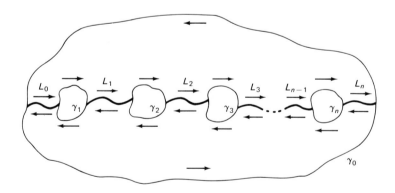

Figure 2.6

A multiply connected domain.

2.3 THE FUNDAMENTAL THEOREM OF INTEGRATION

in opposite directions. Thus the integrals on the arcs L_k cancel out and

$$0 = \int_{\gamma_0 - \sum_{k=1}^{n} \gamma_k} f(z)\, dz = \int_{\gamma_0} f(z)\, dz - \sum_{k=1}^{n} \int_{\gamma_k} f(z)\, dz.$$

Example 1 Since $\cos z$ is entire, has antiderivative $\sin z$, and $\mathscr{C}$ is simply connected, we have

$$\int_{-i}^{i} \cos z\, dz = \sin z \Big|_{-i}^{i} = 2 \sin i = 2i \sinh(1),$$

and along any pwd closed curve γ

$$\int_{\gamma} \cos z\, dz = 0.$$

Example 2 The function $1/z$, analytic in $z \neq 0$, has antiderivative $\log z$. In this case care must be taken to specify the domain G. Suppose G is given by $|\arg z| < \pi$, then for any arc joining $-i$ to i in G

$$\int_{-i}^{i} \frac{dz}{z} = \operatorname{Log} z \Big|_{-i}^{i} = \pi i.$$

On the other hand, if G' is the domain given by $0 < \arg z < 2\pi$, we have

$$\int_{-i}^{i} \frac{dz}{z} = \log z \Big|_{-i}^{i} = \frac{i\pi}{2} - i \frac{3\pi}{2} = -\pi i.$$

Finally, letting $\gamma: z(\theta) = e^{i\theta}$, $-\pi/2 \leq \theta \leq \pi/2$, and $\gamma': z(\theta) = e^{i\theta}$, $\pi/2 \leq \theta \leq 3\pi/2$, we obtain

$$\int_{|z|=1} \frac{dz}{z} = \int_{\gamma} \frac{dz}{z} + \int_{\gamma'} \frac{dz}{z} = \pi i - (-\pi i) = 2\pi i,$$

since γ lies in G and γ' lies in G'.

Example 3†

$$\int_{0}^{\infty} \frac{\sin(x^2)}{x}\, dx = \frac{\pi}{4}.$$

Integrating e^{iz^2}/z along the boundary of $r \leq |z| \leq R$, $0 \leq \arg z \leq \pi/2$ (see Figure 2.7), Cauchy's Theorem yields

$$\int_{r}^{R} \frac{e^{ix^2}}{x}\, dx + i \int_{0}^{\pi/2} e^{i(Re^{i\theta})^2}\, d\theta - \int_{r}^{R} \frac{e^{-iy^2}}{y}\, dy - i \int_{0}^{\pi/2} e^{i(re^{i\theta})^2}\, d\theta = 0. \qquad (3)$$

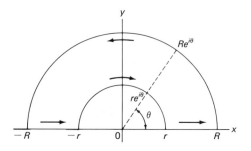

Figure 2.7

But

$$\left| i \int_{0'}^{\pi/2} e^{i(Re^{i\theta})^2} \, d\theta \right| \leq \int_0^{\pi/2} e^{-R^2 \sin 2\theta} \, d\theta$$

$$= 2 \int_0^{\pi/4} e^{-R^2 \sin 2\theta} \, d\theta$$

$$< 2 \int_0^{\pi/4} e^{-4R^2 \theta/\pi} \, d\theta = \frac{\pi}{2R^2} [1 - e^{-R^2}],$$

since $h(\theta) = \sin 2\theta - (4\theta/\pi)$ vanishes at $\theta = 0$, $\pi/4$ and satisfies $h''(\theta) < 0$ for $0 < \theta < \pi/4$, implying that $\sin 2\theta \geq 4\theta/\pi$. Therefore the second integral in (3) vanishes as $R \to \infty$. Given $\varepsilon > 0$, there exists an $r > 0$ such that $|e^{iz^2} - 1| < \varepsilon$ whenever $|z| < r$. Then

$$\left| i \int_0^{\pi/2} e^{i(re^{i\theta})^2} \, d\theta - \frac{i\pi}{2} \right| = \left| i \int_0^{\pi/2} (e^{i(re^{i\theta})^2} - 1) \, d\theta \right| < \varepsilon \frac{\pi}{2},$$

so the last integral in (3) approaches $i\pi/2$ as $r \to \infty$. Adding the first and third integrals in (3) together and letting $R \to \infty$ and $r \to 0$, we have

$$0 = \int_0^\infty \frac{e^{ix^2} - e^{-ix^2}}{x} \, dx - \frac{i\pi}{2} = 2i \int_0^\infty \frac{\sin(x^2)}{x} \, dx - \frac{i\pi}{2}.$$

EXERCISES

1. Evaluate each of these integrals:
   (a) $\int_{-i}^{i} e^{\pi z} \, dz$,
   (b) $\int_{-1}^{i} \sinh(az) \, dz$,
   (c) $\int_1^i (z - 1)^3 \, dz$.

2.3 THE FUNDAMENTAL THEOREM OF INTEGRATION

2. Let $\gamma_1: z(t) = e^{it}$ and $\gamma_2: z(t) = e^{-it}$, $0 \le t \le \pi$. Evaluate the following integrals along each curve:

   (a) $\int \dfrac{dz}{z^2}$,

   (b) $\int \text{Log } z \, dz$,

   (c) $\int \sqrt{z} \, dz$. (*Hint*: Use principal branch of $\sqrt{z}$.)

3. Prove the equalities
$$\int_0^\infty \cos x^2 \, dx = \int_0^\infty \sin x^2 \, dx = \frac{\sqrt{\pi}}{2\sqrt{2}} \qquad \text{Fresnel's integrals,}$$
by applying Cauchy's Theorem to the function $f(z) = e^{-z^2}$ along the boundary of the sector $0 \le |z| \le R$, $0 \le \arg z \le \pi/4$.

4. Show that
$$\int_0^\infty e^{-x^2} \cos(x^2) \, dx = \frac{\sqrt{\pi}}{4} \sqrt{\sqrt{2}+1},$$

$$\int_0^\infty e^{-x^2} \sin(x^2) \, dx = \frac{\sqrt{\pi}}{4} \sqrt{\sqrt{2}-1},$$

by integrating e^{-z^2} along the boundary of the sector $0 \le |z| \le R$, $0 \le \arg z \le \pi/8$.

5. Prove Dirichlet's Integral
$$\int_0^\infty \frac{\sin x}{x} \, dx = \frac{\pi}{2},$$
by integrating $f(z) = e^{iz}/z$ along the boundary of the set $r \le |z| \le R$, $0 \le \arg z \le \pi$. Check your answer by changing variables in Example 3.

6. Suppose $f(z)$ is known to have a continuous derivative $f'(z)$ at all points in the simply connected domain G. Give a proof of Cauchy's Theorem using Green's Theorem for real line integrals
$$\iint_D (p_x + q_y) \, dx \, dy = \int_{\partial D} p \, dy - q \, dx,$$
where D is the inside of a pwd Jordan curve ∂D lying in G, and p and q are continuous on $D \cup \partial D$, and have continuous partial derivatives in x and y on D.

7. Using Green's Theorem as stated in Exercise 6 show that
   (a) $\int_{\partial D} x \, dz = iA$,
   (b) $\int_{\partial D} y \, dz = -A$,
   (c) $\int_{\partial D} \bar{z} \, dz = 2iA$,

   where A equals the area of D.

2.4 THE CAUCHY INTEGRAL FORMULA

We next prove the surprising result that the values of an analytic function inside a pwd Jordan curve are completely determined by its values on the curve:

Cauchy Integral Formula Let $f(z)$ be analytic on a simply connected domain containing the pwd Jordan curve γ. Then

$$f(\zeta) = \frac{1}{2\pi i} \int_\gamma \frac{f(z)}{z - \zeta} \, dz,$$

for all points ζ inside γ.

Proof Fix ζ, then given $\varepsilon > 0$ there exists a closed disk $|z - \zeta| \le r$ lying inside γ for which $|f(z) - f(\zeta)| < \varepsilon$. Since $f(z)/(z - \zeta)$ is analytic on a domain containing the points on and inside γ satisfying $|z - \zeta| \ge r$, the Cauchy Theorem on multiply connected domains implies

$$\frac{1}{2\pi i} \int_\gamma \frac{f(z)}{z - \zeta} \, dz = \frac{1}{2\pi i} \int_{|z - \zeta| = r} \frac{f(z)}{z - \zeta} \, dz.$$

But

$$\int_{|z - \zeta| = r} \frac{f(z)}{z - \zeta} \, dz = f(\zeta) \int_{|z - \zeta| = r} \frac{dz}{z - \zeta} + \int_{|z - \zeta| = r} \frac{f(z) - f(\zeta)}{z - \zeta} \, dz.$$

By Example 1 or Exercise 5 of Section 2.1 the first integral on the right-hand side equals $2\pi i$, so

$$\left| \int_{|z - \zeta| = r} \frac{f(z)}{z - \zeta} \, dz - 2\pi i f(\zeta) \right| \le \int_{|z - \zeta| = r} \frac{|f(z) - f(\zeta)|}{|z - \zeta|} \, |dz| < 2\pi\varepsilon.$$

Since ε can be chosen arbitrarily close to 0, the proof is complete.

2.4 THE CAUCHY INTEGRAL FORMULA

Roughly, Cauchy's Integral Formula says that if γ is deformed into a very small circle around ζ, then $f(z)$ approaches $f(\zeta)$ and

$$\int_\gamma \frac{f(z)}{z - \zeta} dz = \int_{|z - \zeta| = r} \frac{f(\zeta)}{z - \zeta} dz = 2\pi i f(\zeta)$$

(see Figure 2.8).

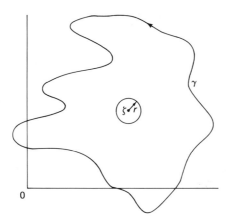

Figure 2.8

Cauchy's Integral Formula.

Example 1 We shall consider the integral

$$\int_\gamma \frac{\cos z}{z^3 + z} dz$$

over three different curves.

(a) $\gamma: |z| = 2$. Decomposing the integral by partial fractions we obtain

$$\int_\gamma \frac{\cos z}{z^3 + z} dz = \int_\gamma \frac{\cos z}{z} dz - \frac{1}{2} \int_\gamma \frac{\cos z}{z + i} dz - \frac{1}{2} \int_\gamma \frac{\cos z}{z - i} dz$$

$$= 2\pi i \left[\cos(0) - \frac{1}{2} \cos(-i) - \frac{1}{2} \cos(i) \right] = 2\pi i [1 - \cosh(1)].$$

(b) $\gamma: |z| = \frac{1}{2}$. Then $\cos z/(z^2 + 1)$ is analytic on and inside γ so the integral equals $2\pi i$ times its value at $z = 0$, that is,

$$\int_\gamma \frac{\cos z}{z^3 + z} dz = 2\pi i.$$

(c) $\gamma: |z - i/2| = 1$. Since $\cos z/(z + i)$ is analytic on and inside γ, by partial fractions we have

$$\frac{1}{z(z-i)} = i\left[\frac{1}{z} - \frac{1}{z-i}\right],$$

$$\int_\gamma \frac{\cos z}{z^3 + z} dz = 2\pi i\left[i\left(\frac{\cos(0)}{i}\right) - i\left(\frac{\cos i}{2i}\right)\right] = 2\pi i\left[1 - \frac{1}{2}\cosh(1)\right].$$

Of course, all three examples can be done utilizing the partial fraction decomposition in part (a), since the corresponding integrals vanish when the points 0 or $\pm i$ lie outside γ.

We now consider the properties possessed by integrals of the type found in the Cauchy Integral Formula:

Riemann's Theorem Let $g(\zeta)$ be continuous on the pwd arc γ. Then the function

$$F_n(z) = \int_\gamma \frac{g(\zeta) \, d\zeta}{(\zeta - z)^n}, \quad n = 1, 2, 3, \ldots,$$

is analytic at all z in the complement of γ, and its derivative satisfies $F'_n(z) = nF_{n+1}(z)$.

Proof† Select a point z_0 not on γ and a disk $|z - z_0| < \delta$ disjoint from γ. For z in the disk $|z - z_0| < \delta/2$ we have

$$|F_1(z) - F_1(z_0)| = \left|\int_\gamma g(\zeta)\left(\frac{1}{\zeta - z} - \frac{1}{\zeta - z_0}\right) d\zeta\right|$$

$$\leq |z - z_0| \int_\gamma \frac{|g(\zeta)| \, |d\zeta|}{|\zeta - z| \, |\zeta - z_0|}.$$

The arc γ has finite length L; thus is a closed and bounded set of points. A theorem of ordinary calculus states that continuous real-valued functions

2.4 THE CAUCHY INTEGRAL FORMULA

attain a maximum on any closed and bounded set. Thus $|g(\zeta)|$ is bounded by M on γ. Since $|\zeta - z| > \delta/2$ for all ζ on γ,

$$|F_1(z) - F_1(z_0)| \leq \frac{2ML}{\delta^2} |z - z_0|,$$

which proves the continuity of $F_1(z)$ at z_0. Applying this fact to the functions

$$G_n(z) = \int_\gamma \frac{\frac{g(\zeta)}{(\zeta - z_0)}}{(\zeta - z)^n} d\zeta,$$

we find that $G_1(z)$ is continuous at z_0 since $g(\zeta)/(\zeta - z_0)$ is continuous on γ. Then, since the difference quotient of $F_1(z)$ equals $G_1(z)$,

$$F_2(z_0) = G_1(z_0) = \lim_{z \to z_0} G_1(z) = \lim_{z \to z_0} \frac{F_1(z) - F_1(z_0)}{z - z_0} = F_1'(z_0).$$

Suppose it were true that $F_{n-1}'(z) = (n-1)F_n(z)$ [and since $g(\zeta)$ is arbitrary, also that $G_{n-1}'(z) = (n-1)G_n(z)$]. Then

$$\frac{1}{\zeta - z} = \frac{1}{\zeta - z_0} + \frac{z - z_0}{(\zeta - z)(\zeta - z_0)}$$

implies that

$$F_n(z) - F_n(z_0) = [G_{n-1}(z) - G_{n-1}(z_0)] + (z - z_0)G_n(z).$$

Since $G_{n-1}(z)$ is differentiable, it is continuous, and

$$|G_n(z)| = \left| \int_\gamma \frac{g(\zeta) \, d\zeta}{(\zeta - z)^n(\zeta - z_0)} \right| \leq \frac{2^n ML}{\delta^{n+1}},$$

for $|z - z_0| < \delta/2$. By the Triangle Inequality

$$0 \leq \lim_{z \to z_0} |F_n(z) - F_n(z_0)| \leq \frac{2^n ML}{\delta^{n+1}} \lim_{z \to z_0} |z - z_0| = 0,$$

implying that $F_n(z)$ [and hence $G_n(z)$] is continuous at z_0. Thus

$$F_n'(z_0) = \lim_{z \to z_0} \left[\frac{G_{n-1}(z) - G_{n-1}(z_0)}{z - z_0} + G_n(z) \right]$$

$$= G_{n-1}'(z_0) + G_n(z_0)$$

$$= nG_n(z_0) = nF_{n+1}(z_0).$$

The proof now follows by induction.

Riemann's Theorem provides a useful generalization of the Cauchy Integral Formula and the remarkable fact that the derivative of an analytic function is also an analytic function:

Theorem Let $f(z)$ be analytic on a simply connected domain containing the pwd Jordan curve γ. Then, for all points ζ inside γ,

$$f^{(n)}(\zeta) = \frac{n!}{2\pi i}\int_\gamma \frac{f(z)}{(z-\zeta)^{n+1}}\,dz, \qquad n = 0, 1, 2, \ldots.$$

Proof Set $g(z) = f(z)$ in Riemann's Theorem. Then

$$F_1(\zeta) = \int_\gamma \frac{f(z)}{z - \zeta}\,dz = 2\pi i f(\zeta)$$

by the Cauchy Integral Formula for all points ζ inside γ. Applying Riemann's Theorem repeatedly, we have

$$F_{n+1}(\zeta) = \frac{F_n'(\zeta)}{n} = \frac{F_{n-1}''(\zeta)}{n(n-1)} = \cdots = \frac{F_1^{(n)}(\zeta)}{n!} = \frac{2\pi i f^{(n)}(\zeta)}{n!},$$

so

$$f^{(n)}(\zeta) = \frac{n!}{2\pi i}F_{n+1}(\zeta) = \frac{n!}{2\pi i}\int_\gamma \frac{f(z)}{(z-\zeta)^{n+1}}\,dz,$$

and the result follows. Adopting the convention that $f^{(0)} = f$ and $0! = 1$, note that the equation above reduces to the Cauchy Integral Formula.

Corollary If $f(z)$ is analytic on a domain G, then so is its derivative $f'(z)$. Furthermore $f(z)$ possesses derivatives of all orders on G.

Proof Since analyticity need only be proved in a neighborhood of a point, for each ζ we can find a disk $|z - \zeta| \le r$ contained in G. Let γ be the circle $|z - \zeta| = r$. Then $f^{(n)}(\zeta)$ exists for all positive integers n, so $f'(z)$ has a derivative at ζ and is thus analytic.

The next result is the converse to Cauchy's Theorem and is often useful in establishing the analyticity of a function, since integration is frequently easier to justify than differentiation:

2.4 THE CAUCHY INTEGRAL FORMULA

Morera's Theorem If $f(z)$ is continuous on a simply connected domain G and satisfies

$$\int_\gamma f(z)\,dz = 0,$$

for all pwd closed curves γ in G, then $f(z)$ is analytic in G.

Proof Select a point z_0 in G and define

$$F(z) = \int_{z_0}^{z} f(z)\,dz,$$

for all z in G. Since the integral along a closed curve vanishes, $F(z)$ is well defined, as the integral is independent of path. As previously shown we have $F_x = f = -iF_y$, so $F(z)$ is analytic in G. But by the corollary, so is $F'(z) = F_x(z) = f(z)$.

Example 2 We shall consider the integral

$$\int_\gamma \frac{\cos z}{z^2(z-1)}\,dz$$

over three different curves (see Figure 2.9).

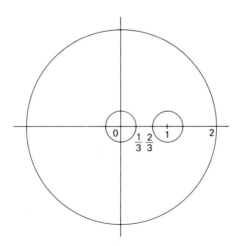

Figure 2.9

(a) $\gamma: |z| = \frac{1}{3}$. In this case $\cos z/(z-1)$ is analytic on and inside γ, so by Cauchy's Formula for derivatives we obtain

$$\int_\gamma \frac{\cos z}{z^2} \frac{}{z-1} dz = 2\pi i \left(\frac{\cos z}{z-1}\right)'\bigg|_{z=0} = -2\pi i.$$

(b) $\gamma: |z-1| = \frac{1}{3}$. Now $z^{-2} \cos z$ is analytic on and inside γ, so the integral equals $2\pi i$ times the value of $z^{-2} \cos z$ at $z = 1$, that is

$$\int_\gamma \frac{z^{-2} \cos z}{z-1} dz = 2\pi i \cos(1).$$

(c) $\gamma: |z| = 2$. By Cauchy's Theorem on multiply connected domains, we may replace γ by the circles in parts (a) and (b). Hence the integral equals $2\pi i[\cos(1) - 1]$. Alternatively, decomposing the integrand by partial fractions, we obtain

$$\int_\gamma \frac{\cos z}{z^2(z-1)} dz = \int_\gamma \cos z \left(\frac{1}{z-1} - \frac{1}{z} - \frac{1}{z^2}\right) dz$$

$$= 2\pi i[\cos(1) - \cos(0) + \sin(0)] = 2\pi i[\cos(1) - 1],$$

by Cauchy's Formula for derivatives.

EXERCISES

1. By decomposing the integrand into partial fractions evaluate the integral

$$\int_\gamma \frac{dz}{(z-a)(z-b)}$$

  (a) if a and b lie inside γ,
  (b) if a lies inside and b outside γ,
  (c) if b lies inside and a outside γ.

2. Let $\gamma: z(t) = 2e^{it} + 1$, $0 \leq t \leq 2\pi$. Evaluate the following integrals:

  (a) $\int_\gamma \frac{e^z}{z} dz$,   (b) $\int_\gamma \frac{\cos z}{z-1} dz$,

  (c) $\int_\gamma \frac{\sin z}{z^2+1} dz$,   (d) $\int_\gamma \frac{\sin z}{z^2 - z} dz.$

2.4 THE CAUCHY INTEGRAL FORMULA

3. Let $\gamma: z(t) = 2e^{it} + 1$, $0 \leq t \leq 2\pi$. Evaluate the following integrals:

   (a) $\int_\gamma \dfrac{e^z}{z^2}\, dz$,

   (b) $\int_\gamma \dfrac{\cos z}{(z-1)^2}\, dz$,

   (c) $\int_\gamma \dfrac{\sin z}{(z^2+1)^2}\, dz$,

   (d) $\int_\gamma \dfrac{\sin z}{(z-1)^3}\, dz$.

4. Let $f(z)$ be analytic in $|z - \zeta| < R$. Prove *Gauss's Mean–Value Theorem*

$$f(\zeta) = \frac{1}{2\pi} \int_0^{2\pi} f(\zeta + re^{i\theta})\, d\theta, \qquad 0 < r < R.$$

5. Compute

$$\int_{|z|=1} \frac{e^{kz^n}}{z}\, dz,$$

for n a positive integer. Then show

$$\int_0^{2\pi} e^{k \cos n\theta} \cos(k \sin n\theta)\, d\theta = 2\pi.$$

6. The Legendre polynomial $P_n(z)$ is defined by

$$P_n(z) = \frac{1}{2^n n!} \frac{d^n}{dz^n} [(z^2 - 1)^n].$$

Using Cauchy's Formula for derivatives, show

$$P_n(z) = \frac{1}{2\pi i} \int_\gamma \frac{(\zeta^2 - 1)^n\, d\zeta}{2^n (\zeta - z)^{n+1}},$$

where z is inside the pwd Jordan curve γ.

7. The Laguerre polynomials $L_n(z)$ are given by

$$L_n(z) = e^z \frac{d^n}{dz^n} (z^n e^{-z}).$$

Show that for all z inside the pwd Jordan curve γ,

$$L_n(z) = \frac{n!}{2\pi i} \int_\gamma \frac{\zeta^n e^{-(\zeta - z)}}{(\zeta - z)^{n+1}}\, d\zeta.$$

8. Suppose $f(z)$ is analytic and satisfies $|f(z)| \leq M$ for $|z - \zeta| \leq r$. Derive *Cauchy's Estimate*

$$|f^{(n)}(\zeta)| \leq Mn!\, r^{-n}.$$

9. Let $f(z)$ be analytic and bounded by M in $|z| \leq R$. Prove
$$|f^{(n)}(z)| \leq \frac{MRn!}{(R-|z|)^{n+1}}, \qquad |z| < R.$$

10. If $f(z)$ is analytic in $|z| < 1$ and $|f(z)| \leq (1-|z|)^{-1}$, show that
$$|f^{(n)}(0)| \leq (n+1)!\left(1 + \frac{1}{n}\right)^n.$$

11. Can an analytic function $f(z)$ satisfy $|f^{(n)}(z)| > n!n^n$, for all positive integers n, at some point z?

12. Prove the following extension of Morera's Theorem: Let $f(z)$ be continuous on the domain G (possibly multiply connected). Suppose that for each ζ in G there is a disk D, containing ζ, in G such that
$$\int_\gamma f(z)\,dz = 0,$$
for all pwd closed curves γ in D. Then $f(z)$ is analytic in G.

2.5 LIOUVILLE'S THEOREM AND THE MAXIMUM PRINCIPLE

In this section we present three useful consequences of the Cauchy Integral Formula and its extension to higher derivatives.

Liouville's Theorem An entire function cannot be bounded unless it is a constant.

Proof Suppose $f(z)$ is entire and bounded by M. Then at any point ζ in $\mathscr{C}$, Cauchy's Estimate (Exercise 8, Section 2.4) yields
$$|f'(\zeta)| = \left|\frac{1}{2\pi i}\int_{|z-\zeta|=r}\frac{f(z)\,dz}{(z-\zeta)^2}\right| \leq \frac{M}{r}.$$
But r can be made arbitrarily large, so $f'(\zeta) = 0$ at all ζ in $\mathscr{C}$. Hence $f(z)$ is constant in $\mathscr{C}$.

Liouville's Theorem yields an easy verification of a very important theorem of elementary algebra which is usually stated without proof:

2.5 LIOUVILLE'S THEOREM AND THE MAXIMUM PRINCIPLE

Fundamental Theorem of Algebra Every polynomial of degree greater than zero has a root.

Proof Suppose $P(z) = a_n z^n + a_{n-1} z^{n-1} + \cdots + a_1 z + a_0$ is not zero for any value z. Then the function $f(z) = 1/P(z)$ is entire. Furthermore $|f(z)|$ approaches zero as $|z|$ tends to infinity, so $|f(z)|$ is bounded for all z. By Liouville's Theorem $f(z)$ and consequently $P(z)$ are constant, contradicting the hypothesis that $n > 0$. Thus $P(z)$ has at least one root.

To show $P(z)$ actually has n roots (including multiple roots) observe that by the Fundamental Theorem it has at least one root, say, ζ_0. Thus $z - \zeta_0$ is a factor of $P(z)$, implying $P(z) = (z - \zeta_0)Q(z)$, where $Q(z)$ has degree $n - 1$. If $n - 1 > 0$, then $Q(z)$ has a root. Continuing in this fashion one may extract n factors of $P(z)$, thus $P(z)$ has exactly n roots.

We next prove one of the most useful theorems in the theory of analytic functions:

Maximum Principle If $f(z)$ is analytic and nonconstant in a domain G, then $|f(z)|$ has no maximum in G.

Proof Suppose there is a point z_0 in G satisfying $|f(z)| \leq |f(z_0)|$ for all z in G. Since z_0 is an interior point there exists a number $r > 0$ such that $|z - z_0| \leq r$ lies in G. Then by the Cauchy Integral Formula (or Exercise 4, Section 2.4)

$$f(z_0) = \frac{1}{2\pi i} \int_{|z-z_0|=r} \frac{f(z)}{z - z_0} dz = \frac{1}{2\pi} \int_0^{2\pi} f(z_0 + re^{it}) dt,$$

that is, the value at the center of the circle equals the integral average of its values on the circle. By assumption $|f(z_0 + re^{it})| \leq |f(z_0)|$, and if strict inequality holds for some value t, it must hold, by the continuity of $|f(z)|$, on an arc of the circle. But then

$$|f(z_0)| \leq \frac{1}{2\pi} \int_0^{2\pi} |f(z_0 + re^{it})| dt < \frac{1}{2\pi} \int_0^{2\pi} |f(z_0)| dt = |f(z_0)|,$$

a contradiction. So $|f(z_0 + re^{it})| = |f(z_0)|$ for $0 \leq t \leq 2\pi$, and since the procedure holds on all circles $|z - z_0| = s$, $0 < s \leq r$, $|f(z)|$ is constant on the disk $|z - z_0| \leq r$. Let S be the set of all points z in G satisfying $|f(z)| = |f(z_0)|$. The argument above shows each such point is an interior point of S, so S is open. But any point in $T = G - S$ is also an interior point by the continuity of $|f(z)|$. Neither T nor S contains a boundary point of the other, as both are open. Since G is connected, T must be empty. Hence $S = G$ and, by the theorem on page 17, $f(z)$ is constant in G, contradicting the hypothesis. Thus $|f(z)|$ has no maximum in G, and the proof is complete.

Denote by $\bar{G}$ the set consisting of G together with its boundary. Since the exterior of G is open, $\bar{G}$ is closed. We can now reformulate the Maximum Principle in the following way:

Corollary Let $f(z)$ be analytic on a bounded domain G and continuous on $\bar{G}$. Then $|f(z)|$ attains its maximum on the boundary of G.

Proof Since $\bar{G}$ is closed and bounded, as in the proof of Riemann's Theorem (Section 2.4), $|f(z)|$ attains a maximum somewhere on $\bar{G}$. By the Maximum Principle, it cannot be in G, so it must be on the boundary of G.

EXERCISES

1. Prove that an entire function satisfying $|f(z)| < |z|^n$ for some n and all sufficiently large $|z|$ must be a polynomial. [*Hint*: Apply the inequalities in Exercise 9, Section 2.4 to either $f^{(n+1)}(z)$ or $f^{(n)}(z)$.]
2. Let $f(z)$ be analytic in $|z| < 1$ and satisfy $f(0) = 0$. Define $F(z) = f(z)/z$ for all z in $0 < |z| < 1$. What value can be given to $F(0)$ in order to make $F(z)$ analytic in $|z| < 1$? (*Hint*: Apply Riemann's Theorem to $F(z)$ on $|z| = r < 1$. Then show the resulting function, analytic in $|z| < r$, coincides with F on $0 < |z| < r$. Use partial fractions.)
3. Using the results in the exercise above and the Maximum Principle prove *Schwarz's Lemma*: Let $f(z)$ be analytic for $|z| < 1$ and satisfy the conditions $f(0) = 0$ and $|f(z)| \le 1$. Then $|f(z)| \le |z|$ and $|f'(0)| \le 1$, with equality only if $f(z) = e^{i\theta}z$ for some fixed real θ.
4. Show that in Schwarz's Lemma $|f(z)| \le 1$ for $|z| < 1$ implies that $|f'(0)| \le 1$ regardless of the value of $f(0)$.
5. Prove the *Minimum Principle*: Let $f(z)$ be analytic on a bounded domain G, continuous and nonzero on $\bar{G}$. Then $|f(z)|$ attains its minimum on the boundary of G. [*Hint*: Consider the function $1/f(z)$.]
6. Give an example to show why the nonzero condition is necessary for the validity of the Minimum Principle.
7. Let $f(z)$ be analytic and nonconstant in $|z| < R$ and denote by $M(r)$ the maximum of $|f(z)|$ on $|z| = r$. Prove that $M(r)$ is strictly increasing for $0 \le r < R$.

NOTES

8. If $f(z)$ is analytic and nonconstant in the bounded domain G, continuous in $\bar{G}$, and has constant absolute value on the boundary of G, then it must have at least one zero in G.

9.† Prove the *Three-Circles Theorem*: If $f(z)$ is analytic on a domain containing the annulus $0 < r_1 \leq |z| \leq r_2$ and satisfies the inequalities $|f(z)| \leq M_1$ on $|z| = r_1$ and $|f(z)| \leq M_2$ on $|z| = r_2$, then the maximum of $|f(z)|$ on $|z| = r$, $r_1 \leq r \leq r_2$, is at most equal to

$$M_1^{(\log r_2/r)/(\log r_2/r_1)} \cdot M_2^{(\log r/r_1)/(\log r_2/r_1)}.$$

NOTES

Section 2.1 The most accessible proof of the Jordan Curve Theorem is found in [W, p. 30].

Section 2.2 The Cauchy–Goursat Theorem may be shown to hold on R with weaker hypotheses. [V, p. 76] proves it is valid for $f(z)$ analytic inside ∂R and continuous on R, and [A, p. 111] allows us to omit the requirement of analyticity at finitely many interior points $z_1, \ldots, z_n$, provided that $(z - z_k)f(z) \to 0$ as $z \to z_k$, $k = 1, \ldots, n$. In fact denumerably many points satisfying the condition may be omitted provided their accumulation points lie on the boundary.

Section 2.3 Verification that $\gamma - \gamma'$ consists of the boundaries of finitely many rectangles may be found in [A, pp. 141–143].

Section 2.4 Further generalizations of Cauchy's Theorem may be found in [A, p. 144] and [Ho, pp. 3, 26]. Riemann's Theorem holds whenever $\int_\gamma |g(\zeta)| \, |d\zeta| < \infty$. The proof, using this weaker hypothesis, is essentially the same. A proof of analyticity of the derivative independent of integration is given by [W, p. 77].

Section 2.5 Generalizations of Schwarz's Lemma may be found in [A, p. 136].

Chapter 3 | INFINITE SERIES

3.1 TAYLOR SERIES

Definition An infinite series of complex numbers
$$a_1 + a_2 + \cdots + a_n + \cdots$$
converges to the sum A if the partial sums
$$S_n = a_1 + a_2 + \cdots + a_n$$
satisfy $S_n \to A$ as $n \to \infty$, and in this case we write $\sum_1^\infty a_n = A$. Otherwise we

3.1 TAYLOR SERIES

say the series *diverges*. A series with the property that the absolute values of its terms form a convergent series is said to be *absolutely convergent*.

Since $a_{n+1} = S_{n+1} - S_n$, if the series converges, we have

$$\lim_{n \to \infty} a_{n+1} = \lim_{n \to \infty} (S_{n+1} - S_n) = A - A = 0.$$

Thus the general term in a convergent series tends to zero. This condition is necessary but of course not sufficient.

An absolutely convergent series must converge. The proof is left as an exercise.

Frequently we are interested in infinite series of functions defined on a domain G

$$\sum_{n=1}^{\infty} f_n(z) = f_1(z) + f_2(z) + \cdots + f_n(z) + \cdots.$$

The series is said to converge on G, if it converges for each z_0 in G. We write

$$f(z) = \sum_{1}^{\infty} f_n(z)$$

and call $f(z)$ the *sum* of the series.

We show that every analytic function can be expressed as a convergent Taylor series:

Taylor's Theorem Let $f(z)$ be analytic in the domain G containing the point z_0. Then the representation

$$f(z) = f(z_0) + \frac{f'(z_0)}{1!}(z - z_0) + \cdots + \frac{f^{(n)}(z_0)}{n!}(z - z_0)^n + \cdots$$

holds in all disks $|z - z_0| < r$ contained in G.

Proof Using Cauchy's Formula for derivatives, and the identity

$$\frac{1}{1-a} = 1 + a + a_2 + \cdots + a^{n-1} + \frac{a^n}{1-a}, \qquad a \ne 1,$$

we have for any closed disk $|\zeta - z_0| \leq r$ contained in G

$$f(z) = \frac{1}{2\pi i} \int_{|\zeta - z_0| = r} \frac{f(\zeta)}{\zeta - z} d\zeta$$

$$= \frac{1}{2\pi i} \int_{|\zeta - z_0| = r} \frac{f(\zeta)}{\zeta - z_0} \left[\frac{1}{1 - \frac{z - z_0}{\zeta - z_0}} \right] d\zeta$$

$$= \frac{1}{2\pi i} \int_{|\zeta - z_0| = r} \frac{f(\zeta)}{\zeta - z_0}$$
$$\times \left[1 + \frac{z - z_0}{\zeta - z_0} + \cdots + \left(\frac{z - z_0}{\zeta - z_0}\right)^{n-1} + \frac{(z - z_0)^n}{(\zeta - z)(\zeta - z_0)^{n-1}} \right] d\zeta$$

$$= f(z_0) + (z - z_0) \frac{f'(z_0)}{1!} + \cdots + (z - z_0)^{n-1} \frac{f^{(n-1)}(z_0)}{(n-1)!} + R_n,$$

where

$$R_n = \frac{(z - z_0)^n}{2\pi i} \int_{|\zeta - z_0| = r} \frac{f(\zeta)}{(\zeta - z)(\zeta - z_0)^n} d\zeta$$

(see Figure 3.1). If we select z inside $|\zeta - z_0| = r$, let $|z - z_0| = \rho$, and observe that $|\zeta - z| \geq r - \rho$ for all ζ on $|\zeta - z_0| = r$, we have

$$|R_n| \leq \frac{\rho^n}{2\pi} \frac{2\pi r M}{(r - \rho) r^n} = \frac{rM}{r - \rho} \left(\frac{\rho}{r}\right)^n,$$

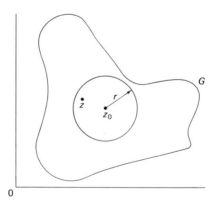

Figure 3.1
The disk $|\zeta - z_0| < r$ is contained in G.

3.1 TAYLOR SERIES

with M the maximum of $|f(\zeta)|$ on $|\zeta - z_0| = r$ (see Figure 3.1). But $\rho/r < 1$, hence $R_n \to 0$ as $n \to \infty$. Therefore $f(z)$ is represented by the Taylor series for all such z.

We notice that this theorem allows us to obtain Taylor series for analytic functions in the same manner as was done in ordinary calculus. For example, if $f(z) = e^z$, since $f^{(n)}(z) = e^z$ and $f^{(n)}(0) = 1$, we have the *Maclaurin series*

$$e^z = \sum_{n=0}^{\infty} \frac{z^n}{n!}, \qquad |z| < \infty,$$

valid for all z in $\mathscr{C}$ since $f(z)$ is entire.

The next two theorems are useful consequences of Taylor's Theorem:

Theorem If $f(z)$ is analytic in a domain G containing the point z_0, and $f^{(n)}(z_0) = 0$ for $n = 1, 2, \ldots$, then $f(z)$ is constant in G.

Proof Let $g(z) = f(z) - f(z_0)$, then g is analytic in G and satisfies $g^{(n)}(z_0) = 0$ for $n = 0, 1, 2, \ldots$. By the proof of Taylor's Theorem, $g(z) = R_n$ which vanishes as $n \to \infty$, provided z lies in a disk $|\zeta - z_0| < r$ contained in G. Thus $g^{(n)}(z) = 0$, $n = 0, 1, 2, \ldots$, for all z in this disk. Let S be the set of all points z in G at which $g^{(n)}(z) = 0$, $n = 0, 1, 2, \ldots$, and let $T = G - S$. By the argument above, S is open, as all its points are interior points. If z_1 is in T, there is an integer $n \geq 0$ for which $g^{(n)}(z_1) \neq 0$. Thus in a disk centered on z_1 lying in G the Taylor series of $g(z)$ does not vanish, implying that z_1 is an interior point of T. Thus T is open. Neither T nor S contains a boundary point of the other as both are open. Since G is connected, T must be empty, hence $g(z) = f(z) - f(z_0) = 0$ for all z in G.

Notice that this theorem implies that if a nonconstant function $f(z)$ analytic in a domain G vanishes at a point z_0 in G, then there is a positive integer n for which $f^{(n)}(z_0) \neq 0$. The least such integer determines the *order of the zero of f at z_0*, and allows us to write

$$f(z) = (z - z_0)^n f_n(z), \qquad f_n(z) = \frac{1}{2\pi i} \int_{|\zeta - z_0| = r} \frac{\dfrac{f(\zeta)}{(\zeta - z_0)^n}}{\zeta - z} \, d\zeta,$$

with $f_n(z)$ analytic inside the disk $|\zeta - z_0| \le r$ contained in G by Riemann's Theorem. Furthermore

$$f_n(z_0) = \frac{1}{2\pi i} \int_{|\zeta - z_0| = r} \frac{f(\zeta)}{(\zeta - z_0)^{n+1}} d\zeta = \frac{f^{(n)}(z_0)}{n!} \ne 0.$$

Thus there is an ε-neighborhood of z_0 lying in G on which $f_n(z_0)$ does not vanish, since f_n is continuous. This shows that z_0 is the only zero of f in $|z - z_0| < \varepsilon$. We have proved:

Theorem The zeros of a nonconstant analytic function are isolated.

Example 1 To obtain the Maclaurin series of $f(z) = (1 - z)^{-2}$, note that

$$f^{(n)}(z) = (n + 1)!(1 - z)^{-(n+2)}, \quad n = 0, 1, 2, \ldots.$$

Hence we have $f^{(n)}(0) = (n + 1)!$ and

$$\frac{1}{(1 - z)^2} = \sum_{n=0}^{\infty} (n + 1)z^n, \quad |z| < 1,$$

since $f(z)$ is not analytic at $z = 1$. The Taylor series of $f(z)$ centered at the point $z_0 = -1$ is

$$\frac{1}{(1 - z)^2} = \sum_{n=0}^{\infty} \frac{(n + 1)}{2^{n+2}} (z + 1)^n, \quad |z| < 2,$$

since $f^{(n)}(-1) = (n + 1)!/2^{n+2}$.

Example 2 To find the order of the zero of $f(z) = 2z(e^z - 1)$ at $z = 0$, calculate $f^{(n)}(0)$ for $n = 0, 1, 2, \ldots$. As

$$f'(z) = 2ze^z + 2(e^z - 1), \quad f''(z) = 2ze^z + 4e^z,$$

we find $f''(0) = 4$, so the order is 2. This is also clear from the Maclaurin series of $f(z)$

$$2z(e^z - 1) = 2z \sum_{n=1}^{\infty} \frac{z^n}{n!} = 2z^2 \left(1 + \frac{z}{2!} + \frac{z^2}{3!} + \cdots\right).$$

EXERCISES

1. Give an example of a divergent series whose general term tends to zero.
2. Prove that an absolutely convergent series must converge.

3.1 TAYLOR SERIES

3. Obtain the following Maclaurin series:

(a) $\sin z = \sum_{n=1}^{\infty} (-1)^{n-1} \frac{z^{2n-1}}{(2n-1)!}$, $\quad |z| < \infty$,

(b) $\cos z = \sum_{n=0}^{\infty} (-1)^n \frac{z^{2n}}{(2n)!}$, $\quad |z| < \infty$,

(c) $\sinh z = \sum_{n=1}^{\infty} \frac{z^{2n-1}}{(2n-1)!}$, $\quad |z| < \infty$,

(d) $\cosh z = \sum_{n=0}^{\infty} \frac{z^{2n}}{(2n)!}$, $\quad |z| < \infty$,

(e) $\frac{1}{1-z} = \sum_{n=0}^{\infty} z^n$, $\quad |z| < 1$.

4. Expand the given functions in a Taylor series about z_0:

(a) $f(z) = \frac{1}{1-z}$, $\quad z_0 = -1$,

(b) $f(z) = \cos z$, $\quad z_0 = \frac{\pi}{2}$,

(c) $f(z) = \frac{1}{z}$, $\quad z_0 = 1$,

(d) $f(z) = \log z$, $\quad z_0 = 1$.

Indicate the largest disk where the representation is valid.

5. Find the order of the zero at $z = 0$ for the functions:

(a) $z^2(\cos z - 1)$, $\qquad$ (b) $6 \sin z^2 + z^2(z^4 - 6)$.

6. If every ε-neighborhood of a point z_0 contains infinitely many points of a set S, we call z_0 an *accumulation point* of the set S. Prove that if two functions analytic on a domain G coincide on a subset of G which has an accumulation point in G, then they coincide everywhere in G.

7. Give an example of two functions which agree at infinitely many points in a domain G, yet are different.

8. Show that if f is a nonconstant analytic function in G, the set of points z satisfying $f(z) = \alpha$, α in $\mathscr{C}$, does not have an accumulation point in G.

3.2 UNIFORM CONVERGENCE OF SERIES

In this section we will prove the converse of Taylor's Theorem, namely that convergent power series are in fact analytic functions in their domain of convergence.

Definition The series $\sum_1^\infty f_n(z)$ is *uniformly convergent* on G if for every $\varepsilon > 0$ there exists a positive number K such that

$$\left| f(z) - \sum_{n=1}^k f_n(z) \right| < \varepsilon,$$

for any $k > K$ and z in G.

Uniform convergence differs from convergence in that for convergence we need only show the existence of a positive *function* $K(z)$, such that for each z_0 in G,

$$\left| f(z_0) - \sum_{n=1}^k f_n(z_0) \right| < \varepsilon,$$

whenever $k > K(z_0)$. The importance of the concept of uniform convergence comes from the following result:

Weierstrass's Theorem The sum of a uniformly convergent series of *analytic* functions is analytic and may be differentiated or integrated term-by-term.

Proof Let $f(z) = \sum_1^\infty f_n(z)$ with each $f_n(z)$ analytic in the domain G. Given $\varepsilon > 0$ there is a positive K such that

$$\left| f(z) - \sum_{n=1}^k f_n(z) \right| < \frac{\varepsilon}{3},$$

for $k > K$ and all z in G. For any z_0 in G and fixed k ($> K$) there is a δ such that

$$\left| \sum_{n=1}^k f_n(z) - \sum_{n=1}^k f_n(z_0) \right| < \frac{\varepsilon}{3},$$

3.2 UNIFORM CONVERGENCE OF SERIES

whenever z is in G and $|z - z_0| < \delta$, since the partial sum is continuous. Thus by the Triangle Inequality

$$|f(z) - f(z_0)| \le \left| f(z) - \sum_{n=1}^{k} f_n(z) \right|$$
$$+ \left| \sum_{n=1}^{k} f_n(z) - \sum_{n=1}^{k} f_n(z_0) \right| + \left| \sum_{n=1}^{k} f_n(z_0) - f(z_0) \right|$$
$$< \varepsilon,$$

whenever z lies in G and $|z - z_0| < \delta$. Hence f is continuous in G. By Cauchy's Theorem, for any pwd closed curve γ lying in a disk contained in G,

$$\left| \int_\gamma f(z)\,dz \right| = \left| \int_\gamma f(z)\,dz - \sum_{n=1}^{k} \int_\gamma f_n(z)\,dz \right|$$
$$\le \int_\gamma \left| f(z) - \sum_{n=1}^{k} f_n(z) \right| |dz| < \frac{\varepsilon L}{3} < \infty,$$

where L is the length of γ. Since ε can be made arbitrarily close to zero, the extension of Morera's Theorem (Exercise 12, Section 2.4) holds and $f(z)$ is analytic in G. (If G is simply connected it follows directly from Morera's Theorem.) In particular, the above shows that on any pwd arc γ in G

$$\int_\gamma f(z)\,dz = \sum_{n=1}^{\infty} \int_\gamma f_n(z)\,dz.$$

Furthermore, by Cauchy's Formula for derivatives,

$$\left| f'(z) - \sum_{n=1}^{k} f_n'(z) \right| = \left| \frac{1}{2\pi i} \int_{|\zeta - z_0| = r} \frac{f(\zeta) - \sum_{n=1}^{k} f_n(\zeta)}{(\zeta - z)^2}\,d\zeta \right| < \frac{4\varepsilon}{3r},$$

for all z satisfying $|z - z_0| < r/2$, $k > K$, and $|\zeta - z_0| \le r$ contained in G (see Figure 3.2). Thus the series $\sum f_n'(z)$ converges uniformly to $f'(z)$ on $|z - z_0| < r/2$, and the proof is complete.

Weierstrass's Theorem can be applied to the power series $\sum_1^\infty a_n(z - z_0)^n$, since each term in the series is an entire function. Before proceeding in this direction, it is useful to make some comments about power series. Note that the substitution $\zeta = z - z_0$ transforms the series above into the power series $\sum_1^\infty a_n \zeta^n$, so we shall only consider power series of this last form.

From elementary calculus, recall the concept of the *radius of convergence* R of a power series $\sum_1^\infty r_n x^n$, where the coefficients r_n are real. The number

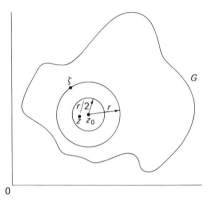

Figure 3.2
$|\zeta - z| > r/2$.

$0 \leq R \leq \infty$ has the property that the series converges absolutely for $|x| < R$ and diverges for $|x| > R$. Then R can be calculated by the formula

$$R = \lim_{n \to \infty} \left| \frac{r_n}{r_{n+1}} \right|,$$

provided the limit exists. Unfortunately for series like

$$2 + x + 2x^2 + x^3 + \cdots + 2x^{2k} + x^{2k+1} + \cdots$$

the ratio of the coefficients $|r_n/r_{n+1}|$ is alternately $\frac{1}{2}$ and 2, so no limit exists. We now give a formula which can always be used in calculating the radius of convergence R of a power series $\sum_1^\infty a_n z^n$, and prove R behaves in the same fashion as it did for real power series.

Hadamard's Formula The radius of convergence R of a power series $\sum_1^\infty a_n z^n$ is given by

$$R^{-1} = \limsup_{n \to \infty} \sqrt[n]{|a_n|} = \lim_{n \to \infty} [\mathrm{lub}(|a_n|^{1/n}, |a_{n+1}|^{1/(n+1)}, \ldots)].$$

The least upper bound (lub) either decreases or remains constant as n increases, so this limit always exists (with infinity a permissible value). Since $2^{1/(2n)} \to 1$, as $n \to \infty$, the series

$$2 + x + 2x^2 + x^3 + \cdots$$

has radius of convergence $R = 1$.

3.2 UNIFORM CONVERGENCE OF SERIES

Abel's Theorem Let R be the radius of convergence of the power series $\sum_1^\infty a_n z^n$. Then

(i) The series converges absolutely in $|z| < R$ and uniformly in $|z| \leq \rho$, $\rho < R$.

(ii) The series diverges in $|z| > R$.

(iii) The sum of the series is analytic in $|z| < R$, and its derivative, obtained by termwise differentiation, has the same radius of convergence.

Proof (i) Let $|z| < r < R$. Then $r^{-1} > R^{-1}$, so the definition of lim sup implies the existence of an integer N such that $|a_n| < r^{-n}$ for all $n \geq N$. Then

$$\sum_{n=N}^\infty |a_n||z|^n < \sum_{n=N}^\infty \left|\frac{z}{r}\right|^n = \frac{\left|\frac{z}{r}\right|^N}{1 - \left|\frac{z}{r}\right|},$$

by the geometric series [Exercise 1(e), Section 3.1], since $|z/r| < 1$. Thus the series converges absolutely in $|z| < r$ for any $r < R$, hence in $|z| < R$. To show uniform convergence select $|z| \leq \rho < r < R$, then by what was done above and the Triangle Inequality

$$\left|\sum_{n=k}^\infty a_n z^n\right| \leq \sum_{n=N}^\infty |a_n||z|^n < \frac{\left(\frac{\rho}{r}\right)^N}{1 - \left(\frac{\rho}{r}\right)},$$

for all $k \geq N$ and $|z| \leq \rho$.

(ii) If $|z| > r > R$, then $r^{-1} < R^{-1}$ and the definition of lim sup yields the existence of infinitely many integers n for which $r^{-n} < |a_n|$. Therefore infinitely many terms of the sequence satisfy $|a_n z^n| > |z/r|^n$, and thus are unbounded.

(iii) That the sum is analytic in $|z| < R$ and its derivative can be obtained by termwise differentiation follows from Weierstrass's Theorem. Set $\sqrt[n]{n} = 1 + c_n$, then by the Binomial Theorem

$$n = (1 + c_n)^n > 1 + \frac{n(n-1)}{2} c_n^2,$$

implying that $c_n^2 < 2/n$, so $c_n \to 0$ as $n \to \infty$. We compute the radius of convergence of the derivative $\sum_1^\infty n a_n z^{n-1}$ as

$$\limsup_{n\to\infty} \sqrt[n]{|a_n|} \leq \limsup_{n\to\infty} \sqrt[n]{n|a_n|}$$
$$\leq \lim_{n\to\infty} \sqrt[n]{n} \cdot \limsup_{n\to\infty} \sqrt[n]{|a_n|} = \limsup_{n\to\infty} \sqrt[n]{|a_n|}.$$

Example 1 Consider the series

$$1 - z^2 + z^4 - z^6 + \cdots.$$

Then $|a_n|$ vanishes for odd n and equals 1 for even n. So $R = 1$, thus the series converges absolutely in $|z| < 1$, uniformly in $|z| \leq r < 1$, and diverges in $|z| > 1$. Moreover, it represents some analytic function in $|z| < 1$. Nothing is said about $|z| = 1$, however, observe that it diverges everywhere on $|z| = 1$ since its general term does not tend to zero. Applying Exercise 1(e), Section 3.1, we find that

$$\frac{1}{1+z^2} = 1 - z^2 + z^4 - z^6 + \cdots, \qquad |z| < 1.$$

Note that the series is analytic only in $|z| < 1$ while the function $(1 + z^2)^{-1}$ is analytic everywhere in $\mathscr{C}$ except at $z = \pm i$. We may integrate the series termwise on any path inside the unit circle obtaining

$$\int_0^z \frac{dz}{1+z^2} = z - \frac{z^3}{3} + \frac{z^5}{5} - \frac{z^7}{7} + \cdots, \qquad |z| < 1.$$

In particular, the function

$$f(z) = \frac{1}{z}\int_0^z \frac{dz}{1+z^2} = 1 - \frac{z^2}{3} + \frac{z^4}{5} - \cdots, \qquad 0 < |z| < 1,$$

$$f(0) = 1,$$

is analytic in $|z| < 1$, illustrating a useful way to show analyticity.

Example 2 To find the radius of convergence of the power series

(a) $\sum_{n=0}^{\infty} \frac{z^n}{n}$, (b) $\sum_{n=0}^{\infty} \frac{z^n}{n!}$, (c) $\sum_{n=0}^{\infty} 2^n z^{n!}$,

note, in (a), that $(1/n)^{1/n} = 1/\sqrt[n]{n}$ tends to 1 as $n \to \infty$, by the proof of Abel's Theorem. Thus $R = 1$ for the series (a). Observe that this result could have been obtained using the ratio formula for the radius of convergence.

3.2 UNIFORM CONVERGENCE OF SERIES

For (b), it is easier to use the ratio formula

$$\left|\frac{r_n}{r_{n+1}}\right| = \frac{(n+1)!}{n!} = n+1 \to \infty, \qquad n \to \infty,$$

and $R = \infty$. The series in (c) is one in which the ratio formula cannot be used, as it contains infinitely many zero coefficients. Here $R = 1$, since we have

$$(2^n)^{1/n!} = e^{\ln 2/(n-1)!} \to e^0 = 1, \qquad n \to \infty,$$

and all the other terms vanish.

Example 3 The Maclaurin series satisfying the differential equation $f''(z) - z^2 f(z) = 9z^3$ with initial conditions $f(0) = 0$, $f'(0) = 1$, can be obtained by differentiating the Taylor series

$$f(z) = a_1 z + a_2 z^2 + a_3 z^3 + \cdots + a_n z^n + \cdots, \qquad a_1 = 1,$$

twice, obtaining

$$f''(z) = 2a_2 + 6a_3 z + 12a_4 z^2 + \cdots + (n+2)(n+1)a_{n+2} z^n + \cdots.$$

Then we have

$$f''(z) - z^2 f(z) = 2a_2 + 6a_3 z + 12a_4 z^2 + \sum_{n=3}^{\infty} [(n+2)(n+1)a_{n+2} - a_{n-2}]z^n$$

$$= 9z^3,$$

so $a_2 = a_3 = a_4 = 0$ and $(n+2)(n+1)a_{n+2} = a_{n-2} + 9\delta_{3n}$, where δ_{jk} equals 1 when $j = k$ and vanishes elsewhere. Thus all coefficients with subscripts not of the form $4k + 1$ vanish and these last satisfy the recursion formula

$$4k(4k+1)a_{4k+1} = a_{4(k-1)+1} + 9\delta_{1k}.$$

Since $(m + j)(n - j) = mn + j(n - m - j) > mn$ whenever $n > m + j$ and $j > 0$, we find

$$\frac{10}{(4k+1)^{4k+1}} < a_{4k+1} = \frac{10}{4 \cdot 5 \cdot 8 \cdot 9 \cdots (4k)(4k+1)} < \frac{10}{[4(4k+1)]^k}.$$

Thus $f(z)$ is entire ($R = \infty$) since $k \to \infty$ implies

$$\frac{10^{1/4k+1}}{(4k+1)}, \quad \frac{10^{1/4k+1}}{[4(4k+1)]^{1/4 + (1/k)}} \to 0.$$

EXERCISES

1. Find the radius of convergence of the series:

(a) $\sum_{n=1}^{\infty} (nz)^n$,

(b) $\sum_{n=1}^{\infty} \frac{n! z^n}{n^n}$,

(c) $\sum_{n=0}^{\infty} z^{2n}$,

(d) $\sum_{n=0}^{\infty} z^{n!}$,

(e) $\sum_{n=0}^{\infty} [2 + (-1)^n]^n z^n$,

(f) $\sum_{n=0}^{\infty} (\cos in) z^n$.

2. If the radius of convergence of the series $\sum_0^{\infty} a_n z^n$ is R ($0 < R < \infty$), find the radius of convergence of the series:

(a) $\sum_{n=0}^{\infty} n^k a_n z^n$,

(b) $\sum_{n=1}^{\infty} n^{-n} a_n z^n$,

(c) $\sum_{n=1}^{\infty} a_n^k z^n$,

(d) $\sum_{n=0}^{\infty} a_n z^{n+k}$,

(e) $\sum_{n=0}^{\infty} a_n^n z^{n^2}$,

(f) $\sum_{n=0}^{\infty} a_n z^{n^2}$.

3. Expand the following functions in power series centered at 0 and find their radius of convergence. Try to avoid using Taylor's Theorem.

(a) $\dfrac{2}{(1-z)^3}$,

(b) $\text{Log}(1 + z)$,

(c) $\displaystyle\int_0^z \frac{\sin z}{z} dz$,

(d) $f(z) = \begin{cases} \dfrac{e^{az} - 1}{z}, & z \neq 0, \\ a, & z = 0. \end{cases}$

4. Find the most general power series (involving two arbitrary constants) satisfying the differential equation $f''(z) + f(z) = 0$. Express the sum in terms of elementary functions.

5. Find a Maclaurin series satisfying the differential equation $f'(z) = 1 + zf(z)$ with the initial condition $f(0) = 0$. What is its radius of convergence?

6. Determine the general Maclaurin series solution of the differential equation $zf''(z) + f'(z) + zf(z) = 0$ and show it is entire. (*Hint:* Prove $\sqrt[n]{n!} \to \infty$ as $n \to \infty$ since e^z is entire.)

3.3 LAURENT SERIES

7. Find the general Maclaurin series solution of the differential equation
$$(1 - z^2)f''(z) - 2zf'(z) + n(n+1)f(z) = 0.$$

8. Suppose $f(z)$ and $g(z)$ are analytic in a neighborhood of z_0 and $f(z_0) = g(z_0) = 0$ while $g'(z_0) \neq 0$. Prove *L'Hospital's Theorem*
$$\lim_{z \to z_0} \frac{f(z)}{g(z)} = \frac{f'(z_0)}{g'(z_0)}.$$

9. Solve Exercise 2, Section 2.5, by the method in this section.
10. Find the sum in $|z| < 1$ of the series
$$\sin \frac{2\pi}{3} + \sin \frac{4\pi}{3} \cdot z + \sin \frac{6\pi}{3} \cdot z^2 + \cdots.$$

11. Prove that if
$$\lim_{n \to \infty} \left| \frac{a_n}{a_{n+1}} \right| = R,$$
then R is the radius of convergence of the series $\sum a_n z^n$.

3.3 LAURENT SERIES

A series of the form
$$a + \frac{a_1}{z} + \frac{a_2}{z^2} + \cdots + \frac{a_n}{z^n} + \cdots$$
can be considered as a power series in the variable $1/z$. If R is its radius of convergence, the series will converge absolutely whenever $|1/z| < R$ or $|z| > 1/R$. The convergence is uniform in every domain $|z| \geq \rho$, $\rho > 1/R$; the series diverges for $|z| < 1/R$ and represents an analytic function in $|z| > 1/R$. Combining a series of the above type with an ordinary power series we obtain a series of the form
$$\sum_{n=-\infty}^{\infty} a_n z^n.$$

Suppose the ordinary part converges in a disk $|z| < R$ and the other part converges in a domain $|z| > r$. If $r < R$, then there is an open annulus where both series converge: $r < |z| < R$ (see Figure 3.3). The series represents an

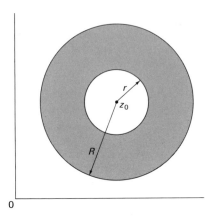

Figure 3.3
Domain of convergence of a Laurent series about z_0.

analytic function in this annulus. Similarly

$$\sum_{n=-\infty}^{\infty} a_n(z-z_0)^n$$

represents an analytic function in the annulus $r < |z - z_0| < R$. We call an expansion of this type a *Laurent series*. We now prove that a function analytic in an annulus $r < |z - z_0| < R$ can be expanded into a Laurent series:

Laurent's Theorem If $f(z)$ is analytic in the annulus $r < |z - z_0| < R$, then it can be uniquely expanded into a Laurent series

$$f(z) = \sum_{n=-\infty}^{\infty} a_n(z-z_0)^n,$$

where

$$a_n = \frac{1}{2\pi i} \int_{|\zeta-z_0|=\rho} \frac{f(\zeta)\, d\zeta}{(\zeta-z_0)^{n+1}}, \qquad n=0,\pm 1, \pm 2, \ldots, \quad r<\rho<R.$$

Proof Let γ_1 and γ_2 denote the circles $|z - z_0| = r + \varepsilon$, $|z - z_0| = R - \varepsilon$, with $0 < \varepsilon < (R-r)/2$, respectively (see Figure 3.4). By Cauchy's Integral Formula

$$f(z) = \frac{1}{2\pi i}\int_{\gamma_2}\frac{f(\zeta)\,d\zeta}{\zeta-z} - \frac{1}{2\pi i}\int_{\gamma_1}\frac{f(\zeta)\,d\zeta}{\zeta-z},$$

3.3 LAURENT SERIES

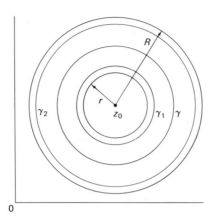

Figure 3.4

for all z satisfying $r + \varepsilon < |z - z_0| < R - \varepsilon$. Riemann's Theorem yields the analyticity of each integral in the complements of the curves γ_1 and γ_2. Proceeding exactly as was done in proving Taylor's Theorem, the first integral becomes

$$\frac{1}{2\pi i} \int_{\gamma_2} \frac{f(\zeta)\, d\zeta}{\zeta - z} = \sum_{n=0}^{\infty} a_n (z - z_0)^n,$$

with

$$a_n = \frac{1}{2\pi i} \int_{\gamma_2} \frac{f(\zeta)\, d\zeta}{(\zeta - z_0)^{n+1}}, \qquad n = 0, 1, 2, \ldots.$$

For the second integral observe that by the geometrical series

$$\frac{-1}{\zeta - z} = \frac{1}{(z - z_0) - (\zeta - z_0)} = \frac{1}{z - z_0} \sum_{n=0}^{\infty} \left(\frac{\zeta - z_0}{z - z_0}\right)^n,$$

since $|\zeta - z_0| < |z - z_0|$ on γ_1. Furthermore the convergence is uniform for all z satisfying $|z - z_0| \geq r + \varepsilon'$, $\varepsilon' > \varepsilon$. By Weierstrass's Theorem we may integrate termwise obtaining

$$-\frac{1}{2\pi i} \int_{\gamma_1} \frac{f(\zeta)}{\zeta - z}\, d\zeta = \sum_{n=0}^{\infty} \left\{ (z - z_0)^{-(n+1)} \cdot \frac{1}{2\pi i} \int_{\gamma_1} f(\zeta)(\zeta - z_0)^n\, d\zeta \right\},$$

thus we have

$$-\frac{1}{2\pi i} \int_{\gamma_1} \frac{f(\zeta)}{\zeta - z}\, d\zeta = \sum_{n=0}^{\infty} a_{-n-1} (z - z_0)^{-n-1},$$

with

$$a_{-n-1} = \frac{1}{2\pi i} \int_{\gamma_1} \frac{f(\zeta)\,d\zeta}{(\zeta - z_0)^{-n}}.$$

Finally, since $f(\zeta)/(\zeta - z_0)^{n+1}$ is analytic inside and on $\gamma - \gamma_1$ or $\gamma_2 - \gamma$, where γ is the circle $|\zeta - z_0| = \rho$, $r < \rho < R$, Cauchy's Theorem implies we may replace γ_1 or γ_2 by γ in the calculation of the coefficients a_n. Noticing that ε can be chosen arbitrarily close to zero yields the desired representation on the annulus $r < |z - z_0| < R$.

The Laurent series representation for a given function is unique, for if $f(z)$ had representations

$$f(z) = \sum_{n=-\infty}^{\infty} a_n(z - z_0)^n, \qquad f(z) = \sum_{n=-\infty}^{\infty} b_n(z - z_0)^n,$$

then multiplying by $(z - z_0)^k$ for any integer k, and integrating along $|z - z_0| = \rho$, would yield by uniform convergence

$$\sum_{n=-\infty}^{\infty} a_n \cdot \int_\gamma (z - z_0)^{n+k}\,dz = \sum_{n=-\infty}^{\infty} b_n \cdot \int_\gamma (z - z_0)^{n+k}\,dz.$$

Since all powers of $z - z_0$ except $(z - z_0)^{-1}$ have analytic antiderivatives in $r < |z - z_0| < R$, their integrals vanish by the Fundamental Theorem of Integration. Thus $2\pi i a_{-k-1} = 2\pi i b_{-k-1}$, implying $a_k = b_k$ for all integers k.

The coefficients a_n are not often obtained by using their integral formulas. We give examples of the techniques employed in avoiding this computation:

Example 1

$$\frac{\cos z}{z^2} = \sum_{n=0}^{\infty} (-1)^n \frac{z^{2n-2}}{(2n)!} = \sum_{n=-1}^{\infty} (-1)^{n+1} \frac{z^{2n}}{(2n+2)!}, \qquad 0 < |z| < \infty,$$

$$e^{1/z^2} = \sum_{n=0}^{\infty} \frac{z^{-2n}}{n!} = \sum_{n=-\infty}^{0} \frac{z^{2n}}{(-n)!}, \qquad 0 < |z|.$$

Example 2 Consider the function $(z^2 - 3z + 2)^{-1}$. It is analytic everywhere except at $z = 1, 2$. In the annulus $1 < |z| < 2$, writing

$$\frac{1}{(z-1)(z-2)} = \frac{1}{z-2} - \frac{1}{z-1},$$

3.3 LAURENT SERIES

we may expand the fractions in the form

$$\frac{1}{z-2} - \frac{1}{z-1} = \frac{-\frac{1}{2}}{1-\frac{z}{2}} - \frac{\frac{1}{z}}{1-\frac{1}{z}}$$

$$= -\frac{1}{2}\sum_{n=0}^{\infty}\left(\frac{z}{2}\right)^n - \frac{1}{z}\sum_{n=0}^{\infty}\left(\frac{1}{z}\right)^n,$$

since $|1/z| < 1$ and $|z/2| < 1$. Thus

$$\frac{1}{(z-1)(z-2)} = -\sum_{n=-\infty}^{-1} z^n - \frac{1}{2}\sum_{n=0}^{\infty}\left(\frac{z}{2}\right)^n.$$

Similar procedures yield

$$\frac{1}{(z-1)(z-2)} = \sum_{n=0}^{\infty}\left(1 - \frac{1}{2^{n+1}}\right)z^n, \qquad |z| < 1,$$

$$= \sum_{n=-\infty}^{-1}\left(\frac{1}{2^{n+1}} - 1\right)z^n, \qquad 2 < |z|,$$

and in $0 < |z-1| < 1$, we obtain

$$\frac{1}{(z-1)(z-2)} = -\frac{1}{z-1} - \frac{1}{1-(z-1)} = -\sum_{n=-1}^{\infty}(z-1)^n, \quad 0 < |z-1| < 1.$$

EXERCISES

1. Represent the function $(z^3 - z)^{-1}$ as a Laurent series in the following domains:
   (a) $0 < |z| < 1$,
   (b) $1 < |z|$,
   (c) $0 < |z-1| < 1$,
   (d) $1 < |z-1| < 2$.

2. Represent the following functions as Laurent series in the domain $0 < |z| < \infty$:
   (a) $ze^{1/z}$,
   (b) $e^{z+1/z}$,
   (c) $\sin z \sin \frac{1}{z}$.

3. Suppose $f(z)$ is analytic and bounded by M in $r < |z - z_0| < R$. Show that the coefficients of its Laurent series satisfy

$$|a_n| \le MR^{-n}, \qquad |a_{-n}| \le Mr^n, \qquad n = 0, 1, 2, \ldots.$$

4. Suppose $r = 0$ in the above exercise. Can $f(z_0)$ be defined in such a way that $f(z)$ is analytic in $|z - z_0| < R$?

5. Bessel's function $J_n(z)$ is defined as the nth coefficient ($n \geq 0$) of the Laurent series of the function

$$e^{(z/2)(\zeta - 1/\zeta)} = \sum_{n=-\infty}^{\infty} J_n(z)\zeta^n.$$

Show that

$$J_n(z) = \frac{1}{\pi} \int_0^\pi \cos(n\theta - z \sin \theta)\, d\theta.$$

3.4 ISOLATED SINGULARITIES

A function $f(z)$ analytic on a domain $0 < |z - z_0| < R$, but not analytic or even necessarily defined at z_0, is said to have an *isolated singularity* at z_0. These singularities are classified in three categories:

(i) *Removable singularities* are those where it is possible to assign a complex number to $f(z_0)$ in such a way that $f(z)$ becomes analytic in $|z - z_0| < R$. In this case it is necessary that $f(z) \to A$ ($\neq \infty$) as $z \to z_0$. But if so, $f(z)$ is analytic on $0 < |z - z_0| < R/2$ and continuous on $|z - z_0| \leq R/2$, hence the Maximum Principle implies that $f(z)$ is bounded on $|z - z_0| \leq R/2$. By Exercise 4, Section 3.3, the Laurent series of $f(z)$ is a convergent Taylor series, thus an analytic function in $|z - z_0| < R$. Therefore the existence of a limit is necessary and sufficient to guarantee the singularity is removable.

(ii) *Poles* occur whenever $f(z) \to \infty$ as $z \to z_0$. In this case observe the function $g(z) = 1/f(z)$ has a removable singularity at z_0 with $g(z_0) = 0$, and z_0 is an isolated zero of $g(z)$ since $f(z)$ is nonzero and finite in a set $0 < |z - z_0| < r \leq R$. If n is the order of the zero of $g(z)$ at z_0, $g(z) = (z - z_0)^n g_n(z)$, then $f(z) = (z - z_0)^{-n} f_n(z)$, where $f_n(z) = 1/g_n(z)$ is analytic in $|z - z_0| < r$. We call n the *order of the pole of f at z_0*. Moreover, the Laurent series of $f(z)$ centered at z_0 is the product of $(z - z_0)^{-n}$ with the Taylor series of $f_n(z)$ at z_0, thus is of the form

$$\sum_{k=-n}^{\infty} a_k (z - z_0)^k, \qquad a_{-n} \neq 0.$$

Since $g(z)$ cannot have zeros of infinite order, $f(z)$ cannot have poles of infinite order.

3.4 ISOLATED SINGULARITIES

(iii) *Essential singularities* are all isolated singularities that are not removable or poles. In this case $f(z)$ does not have a limit as $z \to z_0$, and infinitely many of the coefficients a_{-n}, $n = 1, 2, 3, \ldots$, of its Laurent series about z_0 do not vanish, since otherwise z_0 is a pole or a removable singularity.

We give some examples to amplify our definitions.

$$\frac{\sin z}{z} = 1 - \frac{z^2}{3!} + \frac{z^4}{5!} - \cdots$$

has a removable singularity at $z = 0$. Thus

$$f(z) = \begin{cases} \dfrac{\sin z}{z}, & z \neq 0, \\ 1, & z = 0, \end{cases}$$

is entire. On the other hand

$$\frac{\cos z}{z} = \frac{1}{z} - \frac{z}{2!} + \frac{z^3}{4!} - \cdots$$

has a pole of order 1 at $z = 0$. Finally

$$e^{1/z} = 1 + \frac{1}{z} + \frac{1}{2! z^2} + \cdots$$

has an essential singularity at $z = 0$.

The concept of isolated singularities also applies to (single-valued) functions $f(z)$ analytic in a neighborhood $R < |z| < \infty$ of ∞. By convention, we classify an isolated singularity at ∞ according as to whether $g(z) = f(1/z)$ has a removable singularity, pole, or essential singularity at $z = 0$.

Singularities need not be isolated. For example, the function

$$f(z) = \left(\sin \frac{1}{z}\right)^{-1}$$

has singularities at $z = (\pi n)^{-1}$ for all positive integers n. Thus $z = 0$ is not an isolated singularity.

Definition A function analytic in a domain G, except for poles, is said to be *meromorphic* in G.

If $f(z)$ and $g(z)$ are analytic in G, and $g(z)$ is not identically zero, then the singularities of the quotient $f(z)/g(z)$ agree with the zeros of $g(z)$. They are poles whenever $f(z)$ is nonzero or has a zero of order less than that of $g(z)$, otherwise they are removable singularities. Extending $f(z)/g(z)$ by continuity over the removable singularities we obtain a meromorphic function in G. For example: $f(z) = \tan z$ is meromorphic in $\mathscr{C}$ with poles at $z = (k + \frac{1}{2})\pi$, $k = 0, \pm 1, \pm 2, \ldots$, and $z = \infty$ is an accumulation point of poles.

The behavior of a function in an ε-neighborhood of an essential singularity is very complicated as is shown by the following result:

Weierstrass–Casorati Theorem An analytic function approaches any given value arbitrarily closely in any ε-neighborhood of an essential singularity.

Proof If the theorem is false, we can find a complex number A and a $\delta > 0$ satisfying $|f(z) - A| > \delta$ whenever $0 < |z - z_0| < \delta < \varepsilon$. Then

$$\left|\frac{f(z) - A}{z - z_0}\right| > \frac{\delta}{|z - z_0|} \to \infty, \quad \text{as} \quad z \to z_0,$$

implies that $g(z) = [f(z) - A]/(z - z_0)$ has a pole at z_0 and is, thus, meromorphic in $|z - z_0| < \delta$. But then so is $f(z) = A + (z - z_0)g(z)$, contradicting the hypothesis that z_0 is an essential singularity.

In fact more can be shown although the proof is difficult:

Picard's Theorem An analytic function assumes every complex number, with possibly one exception, infinitely often in any ε-neighborhood of an essential singularity.

Example 1 We wish to find and classify the singularities of the functions

(a) $f(z) = \dfrac{z}{z^2 + z}$,   (b) $g(z) = e^{-1/z^2}$,   (c) $h(z) = \csc z$.

(a) The singularities occur at the zeros of the denominator: $z = 0, -1$. Since these are simple zeros, and the numerator has a simple zero at $z = 0$, $f(z)$ has a removable singularity at $z = 0$ and a simple pole at $z = -1$.

(b) Notice that $g(z) \to 1$, since $1/z^2 \to 0$, as $z \to \infty$, so $g(z)$ has a removable singularity at $z = \infty$. But

$$g(z) = 1 - \frac{1}{z^2} + \frac{1}{2!}\frac{1}{z^4} - \cdots,$$

is the Laurent series of $g(z)$ centered at $z = 0$, hence $g(z)$ has an essential singularity at $z = 0$.

3.4 ISOLATED SINGULARITIES

(c) Since

$$\sin z = (-1)^k \sin(z - \pi k) = (-1)^k \left[(z - \pi k) - \frac{(z - \pi k)^3}{3!} + \cdots \right],$$

$h(z)$ has simple poles at $z = \pi k$, $k = 0, \pm 1, \pm 2, \ldots$, and an accumulation point of poles at $z = \infty$.

Example 2 There is no function analytic in $|z| < 2$ satisfying the condition $f(1/n) = 1 + (-1)^n$, $n = 1, 2, 3, \ldots$, since it would fail to be continuous at $z = 0$.

EXERCISES

1. For each of the following functions find and classify their singularities:

   (a) $\dfrac{z}{z^3 + z}$,
   (b) $\dfrac{e^z}{1 + z^2}$,
   (c) $z e^{1/z}$,
   (d) $e^{z - 1/z}$,
   (e) $\sin \dfrac{1}{z} + \dfrac{1}{z^2}$,
   (f) $e^{\tan 1/z}$.

2. Construct a function having a removable singularity at $z = -1$, a pole of order 3 at $z = 0$, and an essential singularity at $z = 1$. Then find its Laurent series in $0 < |z| < 1$.

3. Does there exist a function analytic in $|z| < 2$ assuming at the points $z = 1/n$, $n = 1, 2, 3, \ldots$, the values:

   (a) $0, 1, 0, -1, 0, 1, 0, -1, \ldots$,
   (b) $1, 0, \tfrac{1}{3}, 0, \tfrac{1}{5}, 0, \tfrac{1}{7}, 0, \tfrac{1}{9}, 0, \ldots$,
   (c) $1, \tfrac{2}{3}, \tfrac{3}{5}, \tfrac{4}{7}, \tfrac{5}{9}, \tfrac{6}{11}, \tfrac{7}{13}, \tfrac{8}{15}, \ldots$,
   (d) $\tfrac{1}{2}, -\tfrac{1}{2}, \tfrac{1}{3}, -\tfrac{1}{3}, \tfrac{1}{4}, -\tfrac{1}{4}, \tfrac{1}{5}, -\tfrac{1}{5}, \ldots$?

4. Show that the function $f(z) = e^{1/z}$ assumes every value except 0 infinitely often in any ε-neighborhood of $z = 0$.

5. Prove an entire function having a nonessential singularity at ∞ must be a polynomial. What kind of singularity do e^z, $\sin z$, and $\cos z$ have at ∞?

6. Show that a function meromorphic in $\mathcal{M}$ must be the quotient of two polynomials.

7. Prove an entire function which omits the values 0 and 1 is constant. (*Hint*: Use Picard's Theorem.)

3.5† ANALYTIC CONTINUATION

It often happens that the expression $f_0(z)$, such as an infinite series or an integral, defining an analytic function is meaningful only in some limited domain G_0 in the plane. The question arises as to whether or not there is any way of extending the definition of the function so that it becomes analytic on a larger domain. In particular, is it possible to find an expression $f_1(z)$ analytic on a domain G_1 meeting G_0 such that $f_0(z) = f_1(z)$, for all z in $G_0 \cap G_1$? If so, we can extend our function to the domain $G_0 \cup G_1$, and we say that the *elements* (f_0, G_0) and (f_1, G_1) form a *direct analytic continuation* of each other. Any direct analytic continuation of (f_0, G_0) to G_1 is necessarily unique, for two functions analytic on G_1 and agreeing on $G_0 \cap G_1$ must coincide on G_1 (see Exercise 6, Section 3.1).

A procedure for obtaining analytic continuations begins by expanding the given expression into a Taylor series

$$f_0(z) = \sum_{n=0}^{\infty} a_n (z - z_0)^n$$

converging in a disk $|z - z_0| < R_0$ centered at a point z_0 in G_0. If z_1 satisfies $|z_1 - z_0| < R_0$, we can expand f_0 in a power series

$$f_1(z) = \sum_{n=0}^{\infty} b_n (z - z_1)^n, \qquad b_n = \frac{f_0^{(n)}(z_1)}{n!},$$

which converges in a disk $|z - z_1| < R_1$. Certainly $R_1 \geq R_0 - |z_1 - z_0|$. If equality holds, the contact point of the circles $|z - z_0| = R_0$ and $|z - z_1| = R_1$ must be a singularity of the function, since Taylor's Theorem implies the

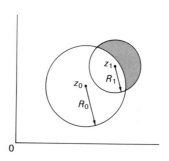

Figure 3.5
Direct analytic continuation.

3.5 ANALYTIC CONTINUATION

existence of a singularity on each circle of convergence. Otherwise, a part of $|z - z_1| < R_1$ lies outside $|z - z_0| < R_0$ and $(f_1, \{|z - z_1| < R_1\})$ is a direct analytic continuation of $(f_0, \{|z - z_0| < R_0\})$, as both series agree on the overlap (see Figure 3.5).

Example 1 The power series

$$f_0(z) = \sum_{n=0}^{\infty} (z - \tfrac{1}{2})^n$$

has radius of convergence $R = 1$, so the domain of convergence G_0 is the disk $|z - \tfrac{1}{2}| < 1$. We can continue (f_0, G_0) to a disk centered at 0 by computing

$$f_0(0) = \sum_{n=0}^{\infty} (-\tfrac{1}{2})^n, \quad f_0'(0) = \sum_{n=1}^{\infty} n(-\tfrac{1}{2})^{n-1}, \ldots,$$

but it is easier to notice that $f_0(z) = (3/2 - z)^{-1}$ in G_0 [see Exercise 3(e), Section 3.1]. Then we have

$$f_1(z) = \frac{2}{3} \frac{1}{1 - \left(\frac{2z}{3}\right)} = \frac{2}{3} \sum_{n=0}^{\infty} \left(\frac{2z}{3}\right)^n, \quad |z| < \frac{3}{2},$$

implying G_1 is the disk $|z| < 3/2$, and $z = 3/2$ is a singularity of the function.

This procedure can be continued, but care must be exercised as a sequence of disks might return to overlap the first, and they might not coincide on the overlap. This occurs when the function is multivalued and the disks have taken us around a branch point of the function and onto a different branch of its Riemann surface (see Figure 3.6). Thus, even if (f_2, G_2) is a direct

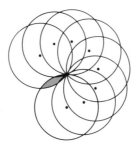

Figure 3.6
An analytic continuation.

analytic continuation of (f_1, G_1), it need not be one for (f_0, G_0), and only a multivalued function will serve to define the extension.

Example 2 Consider the function $f(z) = 1/\sqrt{z}$ at the points $z = e^{\pi i/4}$, $e^{7\pi i/4}$. Using the Binomial Formula, we can obtain the Taylor series expansion about each of these two points

$$\frac{1}{\sqrt{z}} = e^{-\pi i/8} \frac{1}{\sqrt{1 - (1 - ze^{-\pi i/4})}} = e^{-\pi i/8} \sum_{n=0}^{\infty} \frac{(2n)!}{2^{2n}(n!)^2} (1 - ze^{-\pi i/4})^n,$$

$$|z - e^{\pi i/4}| < 1,$$

and

$$\frac{1}{\sqrt{z}} = e^{-7\pi i/8} \frac{1}{\sqrt{1 - (1 - ze^{-7\pi i/4})}} = e^{-7\pi i/8} \sum_{n=0}^{\infty} \frac{(2n)!}{2^{2n}(n!)^2} (1 - ze^{-7\pi i/4})^n,$$

$$|z - e^{7\pi i/4}| < 1.$$

Evaluating the first expression at e^0 and the second at $e^{2\pi i}$ we obtain $e^0 = 1$ and $e^{-\pi i} = -1$, respectively. Note that in the Riemann surface $[\mathscr{C} - \{0\}]^2$, for $f(z) = 1/\sqrt{z}$, the point e^0 does not belong to the disk $|z - e^{7\pi i/4}| < 1$.

Each element of a chain of elements $(f_0, G_0), (f_1, G_1), \ldots, (f_n, G_n)$, such that (f_j, G_j) is a direct analytic continuation of (f_{j-1}, G_{j-1}) is called an *analytic continuation* of the others. Thus, the above procedure can be used to construct analytic continuations, the selection of the centers $z_1, z_2, \ldots, z_n$ determining the values of the function. In particular, if γ is a curve joining z_0 to a point z' not in the disk $|z - z_0| < R_0$, we can construct an analytic continuation consisting of disks of convergence $|z - z_j| < R_j$ of series representations of the function such that z_j follows z_{j-1} in the parametrization of γ. If z' can be reached by a finite chain of such disks, we say we have an *analytic continuation of the function along the curve* γ (see Figure 3.7). Otherwise, we have infinitely many disks whose centers z_j converge to a point z^* on γ, and hence their radii tend to zero. Moreover a singularity of the function must lie on the boundary of each of these disks, and these singularities also tend to z^*. Since every ε-neighborhood of z^* contains a singularity, the function cannot be analytic at z^*. We have proved:

Theorem Along any curve γ which starts in its disk $|z - z_0| < R_0$ of convergence, the power series $\sum_0^{\infty} a_n(z - z_0)^n$ can be continued analytically until one of its singularities is encountered.

3.5 ANALYTIC CONTINUATION

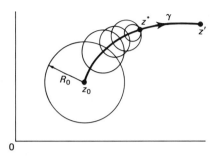

Figure 3.7
Analytic continuation along γ.

Intuitively, if γ and γ' are arcs disjoint except at their two endpoints z_0 and z' such that no singularities lie on or inside the closed curve $\gamma - \gamma'$, then *the result of the analytic continuation is the same for each path*, for the inside could be covered by disks overlapping those in the analytic continuation along the two arcs (see Figure 3.8). We call this result the *Monodromy Theorem*; its proof is complicated.

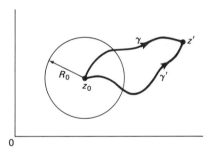

Figure 3.8

EXERCISES

In Exercises 1–3 first find an analytic function agreeing with the given series on its disk of convergence.

1. Expand $\sum_1^\infty z^n/n$ in a neighborhood of $z = \frac{1}{2}$, and determine its radius of convergence.
2. Expand $\sum_{n=1}^\infty z^n$ into a Taylor series in a neighborhood of $z = a$, $|a| < 1$. What is the new series radius of convergence?

3. Show the series

$$\sum_{1}^{\infty} \frac{z^n}{n} \quad \text{and} \quad i\pi + \sum_{1}^{\infty} (-1)^n \frac{(z-2)^n}{n}$$

have no common domain of convergence yet are analytic continuations of each other.

4. Show the function

$$\sum_{n=1}^{\infty} z^{2^n}$$

is analytic in $|z| < 1$, yet cannot be continued outside this set. We call $|z| = 1$ its *natural boundary*. (*Hint*: Since $f(z) = z^2 + z^4 + \cdots + z^{2^n} + f(z^{2^n})$, show the points $\zeta = (1)^{1/2^k}$ satisfy $f(t\zeta) \to \infty$ as $t \to 1^-$.)

5. Show that $|z| = 1$ is a natural boundary for

$$\sum_{n=0}^{\infty} z^{n!}.$$

6. Show the imaginary axis is a natural boundary for the function

$$\sum_{n=0}^{\infty} e^{-n!z}.$$

Where is the function analytic?

7. Find a series representation centered at $z = -1$ for the function

$$f(z) = \int_0^\infty t^2 e^{-zt}\, dt, \quad 0 < t < \infty,$$

analytic in Re $z > 0$. What is its analytic continuation to the whole plane?

8. The Gamma function is defined in the right half plane by means of the integral

$$\Gamma(z) = \int_0^\infty e^{-t} t^{z-1}\, dt, \quad 0 < t < \infty.$$

Prove it satisfies the functional equation $\Gamma(z + 1) = z\Gamma(z)$ and is analytic in Re $z > 0$. Show it has an analytic continuation to the whole plane as a meromorphic function with simple poles at $0, -1, -2, \ldots$.

3.6† RIEMANN SURFACES

The considerations in the last section lead to a precise interpretation of the Riemann surface of a function:

3.6 RIEMANN SURFACES

Definition A global analytic function is a collection $\mathscr{F}$ of elements (f, G) any two of which are analytic continuations of each other by a chain of elements in $\mathscr{F}$.

Example 1 Let G_k be the domain consisting of all points z satisfying $|\arg z - (k\pi/2)| < \pi/2$, for all integers k, and let $f_k(z) = \log z$, for all z in G_k. Then the collection $(f_0, G_0), (f_1, G_1), \ldots, (f_n, G_n), \ldots$ is a general analytic function, as is the collection of elements (f_j, G_j) for all integers j.

Two elements $(f_0, G_0), (f_1, G_1)$ are said to determine the same *branch* of a global analytic function at a point z_0 in $G_0 \cap G_1$ if $f_0(z) = f_1(z)$ in a ε-neighborhood of z_0. Note, however, that it is not necessary that the function elements be direct analytic continuations of each other.

Example 2 If G_k consists of all points z satisfying $|\arg z - (k\pi/2)| < 3\pi/4$, and $f_k(z) = \log z$, for all z in G_k and integers k, then $e^{i\pi/2}$ lies in $G_0 \cap G_2$ and $f_0(z) = f_2(z)$, for all z with $|\arg z - (\pi/2)| < \pi/4$ although (f_0, G_0) and (f_2, G_2) are not direct analytic continuations of each other (see Figure 3.9).

The points on the boundary of the domain of a general analytic function fall into two classes:

(i) points on which the function can be continued analytically (*regular points*), and
(ii) *singularities*.

Singularities may or may not be isolated. If isolated, a singularity is called a *branch point of order* $n - 1$ if all points in an ε-neighborhood of it have n distinct branches. If $n = \infty$, we call it a *logarithmic branch point*.

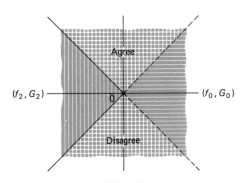

Figure 3.9

As an example, we construct the topological model of the Riemann surface of the function $f(z) = \sqrt{(z-a)(z-b)}$, $a \neq b$. Recall the function $\sqrt{z}$ has a Riemann surface consisting of two copies of $\mathscr{C} - \{0\}$ joined along the negative real axis with 0 and ∞ as branch points of order 1. Hence, the only possible branch points of $f(z)$ are those at which the expression under the radical is either 0 or ∞, so we need only consider $z = a, b, \infty$. The latter may be ruled out, for we may write $f(z) = z\sqrt{(1-a/z)(1-b/z)}$ which behaves like z in a neighborhood of ∞, and ∞ is not a branch point of the function z. The points a and b are branch points of order 1, since letting $z = a + \varepsilon e^{i\theta}$, $0 \leq \theta \leq 4\pi$, we remain on the same branch of $\sqrt{z-b}$ but traverse both branches of $\sqrt{z-a}$, and similarly for $z = b + \varepsilon e^{i\theta}$, $0 \leq \theta \leq 4\pi$, $\varepsilon < |a-b|$. Thus, we may construct the Riemann surface by taking two copies of $\mathscr{M}$ cut along the closed line segment joining a to b, and connect the edges of the cut of one copy of $\mathscr{M}$ to the opposite edges of the other. But there are two line segments in $\mathscr{M}$ joining a to b, one passing through ∞. Since ∞ behaves geometrically like any other point of $\mathscr{M}$, either choice for the branch cut is acceptable (see Figure 3.10).

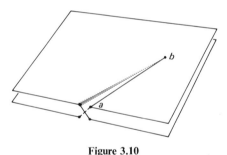

Figure 3.10
Riemann surface of $f(z) = \sqrt{(z-a)(z-b)}$.

Example 3 We construct the Riemann surfaces of the function

$$w = (\log z)^2.$$

Observe that w is the composition of the functions $w = \zeta^2$ and $\zeta = \log z$. Now $\zeta = \sqrt{w}$ has a Riemann surface consisting of two copies of $\mathscr{C} - \{0\}$ joined along the negative real axis with 0 and ∞ as branch points of order 1. The function may be visualized as a mapping of $\mathscr{R}$ onto $[\mathscr{C} - \{0\}]^2$ as illustrated in Figure 3.11. The points $z = 0, \infty$ are logarithmic branch points and $z = 1$ is a regular point.

3.6 RIEMANN SURFACES

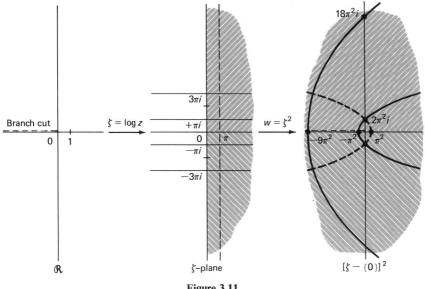

Figure 3.11
$w = (\log z)^2$.

EXERCISES

1. Construct the Riemann surface of the function
$$f(z) = \sqrt{(z-a)(z-b)(z-c)},$$
with a, b, c distinct points, indicating the location of the branch points and their orders.

2. Construct the Riemann surface of the function
$$f(z) = \sqrt{(z-a_1)(z-a_2)\cdots(z-a_n)},$$
for distinct points $a_1, \ldots, a_n$.

3. Construct the Riemann surface of the function
$$f(z) = \sqrt[3]{(z-a)(z-b)(z-c)}.$$

Consider the cases

(i) a, b, c are distinct,
(ii) $a = b \neq c$,
(iii) $a = b = c$.

4. Construct the Riemann surface of the function

$$w = \frac{1}{2}\left(z + \frac{1}{z}\right)$$

and subdivide the z-plane into domains corresponding to half sheets in the Riemann surface.

5. Construct the Riemann surface of the function

$$w = \log \log z,$$

indicating all branch points and their orders, as well as domains corresponding to half sheets in the Riemann surface.

6. Construct the Riemann surface of the function

$$w = \sqrt[n]{\log z}$$

indicating all branch points and their orders, as well as corresponding regions.

7. Prove Pringsheim's Theorem: A power series

$$f(z) = \sum_{n=0}^{\infty} a_n z^n$$

with radius of convergence 1 and nonnegative real coefficients a_n has a singularity at $z = 1$.

NOTES

Important theorems due to Mittag-Leffler and Weierstrass concerning infinite series and products representations for meromorphic functions have been omitted. The reader is urged to study these topics, a development of which may be found in [A, pp. 185–196].

Section 3.4 Two different proofs of Picard's Theorem may be found in [A, p. 297] and [V, p. 144].

Section 3.5 The method indicated for constructing direct analytic continuations, though fine in theory, is rarely useful in practice. The problem lies in computing the coefficients b_n, which, unless additional information is known, are sums of infinite series. Computer techniques will yield good approximations. A proof of the Monodromy Theorem may be found in [A, p. 285].

Section 3.6 There is an extensive amount of literature concerning the theory of Riemann surfaces. An excellent introductory treatment may be found in [Sp].

Chapter 4 | CONTOUR INTEGRATION

4.1 THE RESIDUE THEOREM

We have shown in Section 3.3 that a function $f(z)$, analytic on a domain $0 < |z - z_0| < R$, can be expanded in a Laurent series about z_0. The coefficient

$$a_{-1} = \frac{1}{2\pi i} \int_{|\zeta - z_0| = \rho} f(\zeta)\, d\zeta, \qquad 0 < \rho < R,$$

of this Laurent series is called the *residue* of the function $f(z)$ at z_0 and denoted by $\operatorname{Res}_{z_0} f(z)$.

The next theorem is of fundamental importance in complex analysis and is the core concept in the development of the techniques of this chapter.

Residue Theorem Let $f(z)$ be analytic in a domain G containing the set of all points inside and on a pwd Jordan curve γ except for a finite number of singularities $z_1, \ldots, z_k$ inside γ. Then

$$\int_\gamma f(z)\,dz = 2\pi i \sum_{n=1}^k \mathrm{Res}_{z_n} f(z).$$

Proof We can draw circles $|z - z_n| = r_n\ (>0)$, $n = 1, \ldots, k$, inside γ which are disjoint from each other. By the generalization of Cauchy's Theorem to multiply connected domains

$$\int_\gamma f(z)\,dz = \sum_{n=1}^k \int_{|z-z_n|=r_n/2} f(z)\,dz,$$

and in each domain $0 < |z - z_n| < r_n$, the Laurent series development of $f(z)$ about z_n yields

$$\mathrm{Res}_{z_n} f(z) = a_{-1} = \frac{1}{2\pi i}\int_{|z-z_n|=r_n/2} f(z)\,dz, \qquad n = 1, \ldots, k.$$

Combining these two identities we obtain the desired result.

For this theorem to be useful we need to obtain simple methods for evaluating residues. In particular, we wish to avoid the integration process whenever possible. If the Laurent series is known explicitly then the residue equals a_{-1}. For nonessential singularities we note that a_{-1} vanishes at removable singularities and if z_0 is a pole of order k, then

$$(z - z_0)^k f(z) = \sum_{n=-k}^\infty a_n (z - z_0)^{n+k},$$

so that for $k = 1$,

$$\lim_{z \to z_0} (z - z_0) f(z) = a_{-1},$$

while for $k > 1$,

$$\lim_{z \to z_0} \frac{d^{k-1}}{dz^{k-1}}[(z - z_0)^k f(z)] = (k - 1)!\, a_{-1}.$$

4.1 THE RESIDUE THEOREM

Example 1 We illustrate the above techniques by finding the residue at all singularities in $\mathscr{C}$ of the functions

(a) $f(z) = z^2 \sin \dfrac{1}{z}$, (b) $g(z) = \dfrac{e^z}{z^3 - z^2}$, (c) $h(z) = \dfrac{z}{\sin z}$.

In (a) we know the Laurent series of $f(z)$ centered on $z = 0$,

$$f(z) = z^2 \left(\dfrac{1}{z} - \dfrac{1}{3!z^3} + \dfrac{1}{5!z^5} - \cdots \right) = z - \dfrac{1}{3!z} + \dfrac{1}{5!z^3} - \cdots, \qquad 0 < |z| < \infty,$$

which implies $\operatorname{Res}_0 f(z) = -\tfrac{1}{6}$. For (b), observe $g(z)$ has a simple pole at $z = 1$ and a pole of order 2 at $z = 0$. Thus, we have

$$\operatorname{Res}_1 g(z) = \lim_{z \to 1} (z - 1)g(z) = e,$$

and

$$\operatorname{Res}_0 g(z) = \lim_{z \to 0} [z^2 g(z)]' = \lim_{z \to 0} \dfrac{e^z(z - 2)}{(z - 1)^2} = -2.$$

Finally, $h(z)$ has a removable singularity at $z = 0$ and simple poles at $z = \pi k$, $k = \pm 1, \pm 2, \ldots$ (see Example 1(c), Section 3.4). It follows, since $\sin(z - \pi k) = (-1)^k \sin z$, that the complete solution of problem (c) is given by

$$\operatorname{Res}_{\pi k} h(z) = \lim_{z \to \pi k} \dfrac{(z - \pi k)z}{\sin z} = (-1)^k \pi k, \qquad k = 0, \pm 1, \pm 2, \ldots.$$

We now present, in this and the next four sections, a number of useful techniques for applying the Residue Theorem in the evaluation of definite integrals.

Integrals of the form

$$\int_0^{2\pi} F(\cos \theta, \sin \theta) \, d\theta,$$

where $F(s, t)$ is the quotient of two polynomial functions in s and t, may be transformed into a line integral by the substitution $z = e^{i\theta}$, $0 \le \theta \le 2\pi$, since

$$\cos \theta = \dfrac{1}{2}(e^{i\theta} + e^{-i\theta}) = \dfrac{1}{2}\left(z + \dfrac{1}{z}\right),$$

$$\sin \theta = \dfrac{1}{2i}(e^{i\theta} - e^{-i\theta}) = \dfrac{1}{2i}\left(z - \dfrac{1}{z}\right),$$

and

$$dz = ie^{i\theta} \, d\theta = iz \, d\theta.$$

Theorem

$$\int_0^{2\pi} F(\cos\theta, \sin\theta)\, d\theta = \int_{|z|=1} F\left[\frac{1}{2}\left(z+\frac{1}{z}\right), \frac{1}{2i}\left(z-\frac{1}{z}\right)\right] \frac{dz}{iz}.$$

Example 2

$$\int_0^\pi \frac{d\theta}{a+b\cos\theta} = \frac{\pi}{\sqrt{a^2-b^2}}, \qquad a>b>0.$$

Since $\cos\theta$ takes on the same values on $[\pi, 2\pi]$ as it does on $[0, \pi]$, the integral above equals

$$\frac{1}{2}\int_0^{2\pi} \frac{d\theta}{a+b\cos\theta} = \frac{1}{i}\int_{|z|=1} \frac{dz}{bz^2+2az+b}.$$

Factoring the denominator in $b(z-p)(z-q)$, where

$$p = \frac{-a+\sqrt{a^2-b^2}}{b}, \qquad q = \frac{-a-\sqrt{a^2-b^2}}{b},$$

and observing $pq=1$ and $|q|>a/b>1$, we see that the only singularity of the integrand on the unit disk is at p. Furthermore, it is a pole of order 1, so the residue of the integrand at p equals

$$\lim_{z\to p}\frac{1}{b(z-q)} = \frac{1}{b(p-q)} = \frac{1}{2\sqrt{a^2-b^2}}.$$

The answer now follows by the Residue Theorem.

Example 3

$$\int_0^\pi \frac{d\theta}{(a+b\cos\theta)^2} = \frac{\pi a}{\sqrt{(a^2-b^2)^3}}, \qquad a>b>0.$$

Again, as above, the integral equals

$$\frac{2}{i}\int_{|z|=1}\frac{z\,dz}{(bz^2+2az+b)^2} = \frac{2}{ib^2}\int_{|z|=1}\frac{z\,dz}{(z-p)^2(z-q)^2},$$

with a pole of order 2 at p as the only singularity. The residue at p equals

$$\lim_{z\to p}\left[\frac{z}{(z-q)^2}\right]' = \lim_{z\to p}\frac{-(z+q)}{(z-q)^3} = \frac{-(p+q)}{(p-q)^3} = \frac{ab^2}{4\sqrt{(a^2-b^2)^3}},$$

and the result is now immediate.

4.1 THE RESIDUE THEOREM

EXERCISES

Evaluate the integrals in Exercises 1–9 by the method shown. In Exercises 6–8, n is a nonnegative integer.

1. $\displaystyle\int_0^{\pi/2} \frac{d\theta}{a + \sin^2 \theta} = \frac{\pi}{2\sqrt{a^2 + a}}, \qquad a > 0.$

2. $\displaystyle\int_0^{\pi/2} \frac{d\theta}{(a + \sin^2 \theta)^2} = \frac{\pi(2a + 1)}{4\sqrt{(a^2 + a)^3}}, \qquad a > 0.$

3. $\displaystyle\int_0^{2\pi} \frac{d\theta}{a^2 \cos^2 \theta + b^2 \sin^2 \theta} = \frac{2\pi}{ab}, \qquad a, b > 0.$

4. $\displaystyle\int_0^{2\pi} \frac{d\theta}{(a^2 \cos^2 \theta + b^2 \sin^2 \theta)^2} = \frac{\pi(a^2 + b^2)}{a^3 b^3}, \qquad a, b > 0.$

5. $\displaystyle\int_0^{2\pi} \frac{d\theta}{1 - 2a \cos \theta + a^2} = \begin{cases} \dfrac{2\pi}{1 - a^2}, & \text{if } |a| < 1, \\[6pt] \dfrac{2\pi}{a^2 - 1}, & \text{if } |a| > 1. \end{cases}$

6. $\displaystyle\int_0^{2\pi} \cos^n \theta \, d\theta = \begin{cases} \dfrac{n!\,\pi}{2^{n-1}\left[\left(\dfrac{n}{2}\right)!\right]^2}, & \text{if } n \text{ is even}, \\[10pt] 0, & \text{if } n \text{ is odd}. \end{cases}$

7. $\displaystyle\int_0^{2\pi} (a \cos \theta + b \sin \theta)^n \, d\theta = \begin{cases} \dfrac{n!\,\pi}{2^{n-1}\left[\left(\dfrac{n}{2}\right)!\right]^2} \cdot \sqrt{(a^2 + b^2)^n}, & n \text{ even}, \\[10pt] 0, & n \text{ odd}; \quad a, b \text{ real}. \end{cases}$

8. $\displaystyle\int_0^{2\pi} e^{\cos \theta} \cos(n\theta - \sin \theta) \, d\theta = \frac{2\pi}{n!}.$

9. $\displaystyle\int_0^{2\pi} \cot(\theta + ib) \, d\theta = -2\pi i \operatorname{sign} b, \qquad b \text{ real and nonzero}.$

10. Suppose $P(z)$, $Q(z)$ are polynomials. Show all the residues of the function $[P(z)/Q(z)]'$ vanish.

4.2 EVALUATION OF IMPROPER REAL INTEGRALS

In the last theorem the interval of integration was automatically transformed into a closed curve, allowing us to apply the Residue Theorem. In the next application this is not possible, so instead we replace the given curve by a closed curve such that in the limit the values of the integrals agree.

Theorem Suppose $F(z)$ is the quotient of two polynomials in z such that

  (i) $F(z)$ has no poles on the real axis, and
  (ii) $F(1/z)$ has a zero of order at least two at $z=0$, that is, the degree of the denominator exceeds the degree of the numerator by at least two.

Then

$$\int_{-\infty}^{\infty} F(x) \begin{Bmatrix} \cos ax \\ \sin ax \end{Bmatrix} dx = \begin{Bmatrix} \text{Re} \\ \text{Im} \end{Bmatrix} 2\pi i \sum_{y>0} \text{Res } F(z)e^{iaz}, \qquad a \geq 0,$$

the sum being taken only on the poles of $F(z)$ in the upper half plane.

Proof Let γ be the closed curve obtained by taking the line segment $(-R, R)$ on the real axis followed by the semicircle $z = Re^{i\theta}$, $0 \leq \theta \leq \pi$. Since $F(z)$ is the quotient of polynomials, its poles, and hence those of $F(z)e^{iaz}$, occur only at zeros of the denominator, and thus are finite in number. If R is chosen sufficiently large all poles of $F(z)$ in the upper half plane will lie inside γ. (see Figure 4.1). Then the Residue Theorem implies

$$2\pi i \sum_{y>0} \text{Res } F(z)e^{iaz} = \int_{\gamma} F(z)e^{iaz} \, dz,$$

$$= \int_{-R}^{R} F(x)e^{iax} \, dx + \int_{0}^{\pi} F(Re^{i\theta})e^{iaRe^{i\theta}} iRe^{i\theta} \, d\theta.$$

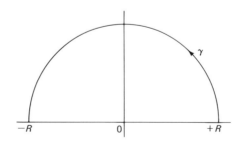

Figure 4.1

4.2 EVALUATION OF IMPROPER REAL INTEGRALS

By (ii) $|z^2 F(z)|$ is bounded by a constant M at all points of the upper half plane not inside γ. Thus

$$\left| \int_0^\pi F(Re^{i\theta}) e^{iaRe^{i\theta}} iRe^{i\theta} \, d\theta \right| \le \frac{M}{R} \int_0^\pi e^{-aR \sin \theta} \, d\theta \le \frac{M\pi}{R},$$

since $e^{-aR \sin \theta} \le 1$. By (ii) and the Comparison Theorem the improper integrals

$$\int_{-\infty}^\infty F(x) \cos ax \, dx, \qquad \int_{-\infty}^\infty F(x) \sin ax \, dx, \qquad a \ge 0,$$

both converge, so letting $R \to \infty$ we have

$$\int_{-\infty}^\infty F(x) e^{iax} \, dx = 2\pi i \sum_{y>0} \operatorname{Res} F(z) e^{iaz}, \qquad a \ge 0,$$

from which the result follows by taking the real and imaginary parts of both sides.

Remark If $a > 0$, condition (ii) can be replaced by:

(ii)′ $F(1/z)$ has a zero of order 1 at $z = 0$.

In this case we cannot use the Comparison Theorem to obtain the convergence of the integral

$$\int_{-\infty}^\infty F(x) e^{iax} \, dx, \qquad a > 0.$$

In fact we must prove

$$\int_{-X_1}^{X_2} F(x) e^{iax} \, dx, \qquad a > 0,$$

has a limit as X_1 and X_2 tend independently to ∞. Let γ be the boundary of the rectangle with vertices at the points $-X_1, X_2, X_2 + iY, -X_1 + iY$, the constants X_1, X_2, Y chosen large enough that the poles of $F(z)$ in the upper half plane lie inside γ (see Figure 4.2). Condition (ii)′ now shows $|zF(z)|$

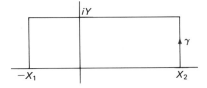

Figure 4.2

bounded by M at all points in $y > 0$ not inside γ. The integral

$$\left| \int_{X_2}^{X_2+iY} F(z)e^{iaz}\, dz \right| \leq M \int_0^Y \frac{e^{-ay}}{|X_2+iy|}\, dy$$

$$\leq \frac{M}{X_2} \int_0^Y e^{-ay}\, dy < \frac{M}{aX_2}.$$

Similarly the integral on the line segment joining $-X_1 + iY$ to $-X_1$ is bounded by M/aX_1 and

$$\left| \int_{X_2+iY}^{-X_1+iY} F(z)e^{iaz}\, dz \right| \leq \frac{Me^{-aY}}{Y} \int_{-X_1}^{X_2} dx = \frac{Me^{-aY}}{Y}(X_1+X_2).$$

Using the Residue Theorem and the Triangle Inequality

$$\left| \int_{-X_1}^{X_2} F(x)e^{iax}\, dx - 2\pi i \sum_{y>0} \operatorname{Res} F(z)e^{iaz} \right| < M\left[\frac{1}{aX_1} + \frac{1}{aX_2} + \frac{e^{-aY}}{Y}(X_1+X_2) \right].$$

First letting $Y \to \infty$ and then letting X_1 and X_2 tend independently to ∞ yields the result.

Example 1

$$\int_0^\infty \frac{\cos ax}{x^2+b^2}\, dx = \frac{\pi e^{-ab}}{2b}, \qquad a \geq 0, \quad b > 0.$$

Here $F(z)$ equals $(z^2+b^2)^{-1}$ with poles at $\pm ib$, and $F(1/z) = z^2/(1+b^2z^2)$ has a zero of order two at 0. As the hypotheses of the theorem are satisfied we have

$$\int_{-\infty}^\infty \frac{\cos ax}{x^2+b^2}\, dx = \operatorname{Re}\left[2\pi i \operatorname{Res}_{ib} \frac{e^{iaz}}{z^2+b^2} \right] = \operatorname{Re}\left[\frac{\pi}{b} e^{-ab} \right],$$

from which the result follows since the integrand is an even function. Note that

$$\int_{-\infty}^\infty \frac{\sin ax}{x^2+b^2}\, dx = 0, \qquad a \geq 0, \quad b > 0.$$

4.2 EVALUATION OF IMPROPER REAL INTEGRALS

Example 2

$$\int_0^\infty \frac{x \sin ax}{x^2 + b^2} dx = \frac{\pi}{2} e^{-ab}, \qquad a > 0, \quad b > 0.$$

Conditions (i) and (ii)' apply to $F(z) = z/(z^2 + b^2)$ so

$$\int_{-\infty}^\infty \frac{x \sin ax}{x^2 + b^2} dx = \text{Im}\left[2\pi i \, \text{Res}_{ib} \frac{ze^{iaz}}{z^2 + b^2}\right] = \pi e^{-ab}$$

the integrand again being an even function.

EXERCISES

Evaluate the integrals below by the method given in this section.

1. $\displaystyle\int_{-\infty}^\infty \frac{x \, dx}{(x^2 + 2x + 2)^2} = -\frac{\pi}{2}.$

2. $\displaystyle\int_{-\infty}^\infty \frac{x^2 \, dx}{(x^2 + 2x + 2)^2} = \pi.$

3. $\displaystyle\int_0^\infty \frac{x^2 \, dx}{(x^2 + a^2)^2} = \frac{\pi}{4a}, \qquad a > 0.$

4. $\displaystyle\int_{-\infty}^\infty \frac{dx}{(x^2 + a^2)(x^2 + b^2)} = \frac{\pi}{ab(a + b)}, \qquad a, b > 0.$

5. $\displaystyle\int_{-\infty}^\infty \frac{dx}{(x^2 + 1)^{n+1}} = \frac{(2n)! \, \pi}{2^{2n}(n!)^2}, \qquad n$ a nonnegative integer.

6. $\displaystyle\int_{-\infty}^\infty \frac{\cos ax \, dx}{(x^2 + b^2)^2} = \frac{\pi(1 + ab)e^{-ab}}{2b^3}, \qquad a \geq 0, \quad b > 0.$

7. $\displaystyle\int_{-\infty}^\infty \frac{x^3 \sin ax \, dx}{(x^2 + b^2)^2} = \frac{\pi}{2}(2 - ab)e^{-ab}, \qquad a, b > 0.$

8. $\displaystyle\int_0^\infty \frac{\cos ax}{x^4 + b^4} dx = \frac{\pi}{2b^3} e^{-(ab)/\sqrt{2}} \sin\left(\frac{ab}{\sqrt{2}} + \frac{\pi}{4}\right), \qquad a \geq 0, \quad b > 0.$

9. $\displaystyle\int_0^\infty \frac{x \sin ax}{x^4 + b^4} dx = \frac{\pi}{2b^2} e^{-(ab)/\sqrt{2}} \sin \frac{ab}{\sqrt{2}}, \qquad a \geq 0, \quad b > 0.$

10. $\displaystyle\int_0^\infty \frac{x^3 \sin ax}{x^4 + b^4} dx = \frac{\pi}{2} e^{-(ab)/\sqrt{2}} \cos \frac{ab}{\sqrt{2}}, \qquad a, b > 0.$

4.3 CONTINUATION

Throughout the discussion in Section 4.2, we assumed the condition that $F(z)$ had no poles on the real axis since otherwise the integral

$$\int_{-\infty}^{\infty} F(x) e^{iax} \, dx, \qquad a > 0.$$

diverges. However, the real or imaginary part of the integral above may converge if $F(z)$ has poles of order 1 coinciding with zeros of $\cos ax$ or $\sin ax$.

Suppose $F(z)$ has a pole of order 1 at $z = 0$ and no other poles on the real axis. Then

$$\int_{-\infty}^{\infty} F(x) \sin ax \, dx, \qquad a > 0,$$

converges. The technique for integration consists of using the boundary γ of the rectangle with vertices at $-X_1, X_2, X_2 + iY, -X_1 + iY$ except that the origin is avoided by following a small semicircle E of radius r in the lower half plane (see Figure 4.3). Assume $X_1, X_2, Y, 1/r$ are chosen sufficiently large

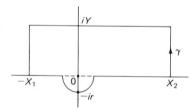

Figure 4.3

that all poles of $F(z)$ not in the lower half plane lie inside γ. Then $F(z)e^{iaz} = (a_{-1}/z) + f(z)$ with $a_{-1} = \text{Res}_0 F(z)e^{iaz}$ and $f(z)$ analytic in a closed ε-neighborhood of $z = 0$. Now on the semicircle E with $r < \varepsilon$,

$$\int_E F(z) e^{iaz} \, dz = i \int_{-\pi}^{0} [a_{-1} + f(re^{i\theta}) re^{i\theta}] \, d\theta$$

$$= \pi i a_{-1} + ir \int_{-\pi}^{0} f(re^{i\theta}) e^{i\theta} \, d\theta.$$

Since $f(z)$ is bounded in $|z| \leq \varepsilon$ by a constant N,

$$\left| ir \int_{-\pi}^{0} f(re^{i\theta}) e^{i\theta} \, d\theta \right| \leq rN\pi,$$

4.3 CONTINUATION

and the second term vanishes as r tends to 0. By the Residue Theorem and the inequalities developed in Section 4.2

$$\int_{-X_1}^{-r} + \int_{E} + \int_{r}^{X_2} F(x)e^{iax}\,dx - 2\pi i \sum_{y \geq 0} \text{Res } F(z)e^{iaz} \to 0$$

by letting $Y \to \infty$ and then X_1 and X_2 tend independently to ∞. Now letting $r \to 0$ we find that

$$\lim_{r \to 0} \int_{-\infty}^{-r} + \int_{r}^{\infty} F(x)e^{iax}\,dx = 2\pi i \left[\sum_{y > 0} \text{Res } F(z)e^{iaz} + \frac{a_{-1}}{2} \right].$$

The limit on the left-hand side of the expression is referred to as the *Cauchy Principal Value* of the integral and written

$$\text{PV} \int_{-\infty}^{\infty} F(x)e^{iax}\,dx = 2\pi i \left[\sum_{y > 0} \text{Res } F(z)e^{iaz} + \frac{a_{-1}}{2} \right].$$

Observe that only half the residue at 0 has been included on the right.

We digress briefly to make some comments about Cauchy Principal Values. Let $f(x)$ be defined on the real line and consider the limits

$$\lim_{R \to \infty} \int_{-R}^{R} f(x)\,dx, \tag{1}$$

$$\lim_{R_1 \to \infty} \int_{-R_1}^{0} f(x)\,dx + \lim_{R_2 \to \infty} \int_{0}^{R_2} f(x)\,dx. \tag{2}$$

If the limit in (1) exists, the improper integral of f is said to *converge in the sense of Cauchy* and we write

$$\text{PV} \int_{-\infty}^{\infty} f(x)\,dx = \lim_{R \to \infty} \int_{-R}^{R} f(x)\,dx.$$

If the limits in (2) exist, we say the improper integral *converges* and set

$$\int_{-\infty}^{\infty} f(x)\,dx = \lim_{R_1 \to \infty} \int_{-R_1}^{0} f(x)\,dx + \lim_{R_2 \to \infty} \int_{0}^{R_2} f(x)\,dx.$$

Note that convergence of the integral implies convergence (to the same value) in the sense of Cauchy, but that an integral may have a principle value without being convergent. For example,

$$\text{PV} \int_{-\infty}^{\infty} x\,dx = \lim_{R \to \infty} \left(\frac{x^2}{2} \bigg|_{-R}^{R} \right) = 0,$$

yet neither limit in (2) exists.

A similar development arises when $f(x)$ is defined in an interval $a \leq x \leq b$ but is unbounded in every neighborhood of a point c, $a < c < b$. The improper integral *converges* provided the right side of the equation

$$\int_a^b f(x)\,dx = \lim_{\varepsilon \to 0} \int_a^{c-\varepsilon} f(x)\,dx + \lim_{\eta \to 0} \int_{c+\eta}^b f(x)\,dx, \qquad \varepsilon > 0, \quad \eta > 0, \qquad (3)$$

exists. Even if these limits fail to exist, the *Cauchy Principal Value* of the integral

$$\text{PV} \int_a^b f(x)\,dx = \lim_{\varepsilon \to 0} \left(\int_a^{c-\varepsilon} f(x)\,dx + \int_{c+\varepsilon}^b f(x)\,dx \right), \qquad \varepsilon > 0, \qquad (4)$$

may exist. For example,

$$\text{PV} \int_{-1}^1 \frac{dx}{x} = \lim_{\varepsilon \to 0} \left(\log \varepsilon + \log \frac{1}{\varepsilon} \right) = 0, \qquad \varepsilon > 0,$$

but neither of the limits in (3) exists.

As before, convergence implies convergence in the sense of Cauchy. Moreover, an improper integral of mixed type may have a Cauchy Principal Value, even though the integral diverges

$$\text{PV} \int_{-\infty}^{\infty} \frac{dx}{x} = \text{PV} \left(\int_{-\infty}^{-1} + \int_1^{\infty} \right) \frac{dx}{x} + \text{PV} \int_{-1}^1 \frac{dx}{x}$$

$$= \lim_{R \to \infty} \left(\int_{-R}^{-1} + \int_1^R \right) \frac{dx}{x} = 0.$$

If $F(z)$ has several poles of order 1 on the real axis coinciding with the zeros of either $\cos ax$ or $\sin ax$, then including as many semicircles as there are poles to γ and treating them as E was treated above yields the following general result:

Theorem Suppose $F(z)$ is the quotient of two polynomials in z such that

(i)′ All poles of $F(z)$ lying on the real axis are of order 1 and coincide with zeros of either $\cos ax$ or $\sin ax$, $a > 0$, and

(ii)′ $F(1/z)$ has a zero of order at least 1 at $z = 0$.

Then

$$\text{PV} \int_{-\infty}^{\infty} F(x) e^{iax}\,dx = 2\pi i \left[\sum_{y > 0} \text{Res } F(z) e^{iaz} + \tfrac{1}{2} \sum_{y = 0} \text{Res } F(z) e^{iaz} \right].$$

4.3 CONTINUATION

Example 1

$$\int_0^\infty \frac{\sin x}{x} \, dx = \frac{\pi}{2}.$$

Since $F(z) = 1/z$, it is clear (i)' and (ii)' hold, so

$$\text{PV} \int_{-\infty}^\infty \frac{e^{ix}}{x} \, dx = \pi i \, \text{Res}_0 \frac{e^{iz}}{z} = \pi i.$$

Equating the imaginary parts yields the desired result since the integrand is an even function.

Integrals containing powers of $\cos ax$ or $\sin ax$ may be evaluated by the same technique:

Example 2

$$\int_0^\infty \frac{\sin^2 x}{x^2} \, dx = \frac{\pi}{2}.$$

Using DeMoivre's Theorem we set $2 \sin^2 x = 1 - \cos 2x$ and obtain

$$\int_{-\infty}^\infty \frac{1 - \cos 2x}{4x^2} \, dx,$$

which converges by the Comparison Theorem of calculus. Integrating $(1 - e^{2iz})/4z^2$ along the curve γ shown in Figure 4.4 we have

$$\int_\gamma \frac{1 - e^{2iz}}{4z^2} \, dz = 2\pi i \, \text{Res}_0 \frac{1 - e^{2iz}}{4z^2} = \pi.$$

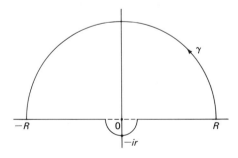

Figure 4.4

The absolute value of the integral along the arc $|z| = R$, $0 \leq \arg z \leq \pi$, is bounded by

$$\frac{1}{4R} \int_0^\pi |1 - e^{2iRe^{i\theta}}| \, d\theta \leq \frac{\pi}{2R},$$

which vanishes as $R \to \infty$. Since

$$\frac{1 - e^{2iz}}{4z^2} = \frac{-i}{2z} + f(z)$$

with $f(z)$ analytic in a closed disk centered at 0 containing E, as in the proof of the theorem we have

$$\left| \int_E \frac{1 - e^{2iz}}{4z^2} \, dz - \frac{\pi}{2} \right| = \left| ir \int_{-\pi}^0 f(re^{i\theta}) e^{i\theta} \, d\theta \right| \leq rN\pi,$$

and this bound vanishes as $r \to 0$. Thus

$$\text{PV} \int_{-\infty}^\infty \frac{1 - e^{2ix}}{4x^2} \, dx = \frac{\pi}{2},$$

and the proof is complete.

EXERCISES

Evaluate the integrals below by the method of this section.

1. $\int_{-\infty}^\infty \frac{\cos \pi x}{4x^2 - 1} \, dx = \frac{-\pi}{2}.$

2. $\int_{-\infty}^\infty \frac{\sin \pi x}{x^5 - x} \, dx = \frac{\pi}{2}(e^{-\pi} - 3).$

3. $\int_{-\infty}^\infty \frac{\sin \pi x \cos \pi x}{2x^2 - x} \, dx = -\pi.$

4. $\int_{-\infty}^\infty \frac{\sin x}{x} \frac{x^2 + a^2}{x^2 + b^2} \, dx = \frac{\pi}{b^2} [a^2 + e^{-b}(b^2 - a^2)], \qquad a, b > 0.$

5. $\int_0^\infty \frac{\sin x}{x(x^2 + b^2)} \, dx = \frac{\pi}{2b^2}(1 - e^{-b}), \qquad b > 0.$

6. $\int_0^\infty \frac{\sin ax}{x(x^2 + b^2)^2} \, dx = \frac{\pi}{2b^4}\left[1 - \frac{e^{-ab}}{2}(ab + 2)\right], \qquad a, b > 0.$

4.4 INTEGRATION OF MULTIVALUED FUNCTIONS

7. $\int_0^\infty \dfrac{\cos ax - \cos bx}{x^2}\,dx = \dfrac{b-a}{2}\pi, \quad a, b \geq 0.$

8. $\int_0^\infty \dfrac{\sin^3 x}{x^3}\,dx = \dfrac{3\pi}{8}.$

9. $\int_{-\infty}^\infty \dfrac{\sin m(x-a)\sin n(x-b)}{x-a \quad x-b}\,dx = \pi\,\dfrac{\sin n(a-b)}{a-b},$

   $m \geq n \geq 0, \quad a, b \text{ real}, \quad a \neq b.$

10. Prove the identity

$$\text{PV}\,\dfrac{1}{2\pi i}\int_{-\infty}^\infty \dfrac{e^{itx}\,dx}{x} = \begin{cases} \tfrac{1}{2}, & t > 0, \\ 0, & t = 0, \\ -\tfrac{1}{2}, & t < 0. \end{cases}$$

Observe that if we add $\tfrac{1}{2}$ to this function, we obtain the "impulse function," often found in engineering books, representing a sudden switch-in of current into an open circuited electric line.

4.4 INTEGRATION OF MULTIVALUED FUNCTIONS

Integrals involving multivalued functions require us to take into account the branch points and branch cuts of the integrand in addition to its isolated singularities. This is due to the fact that, in order to use the Residue Theorem, a domain must be selected in which the integrand is single-valued.

Theorem Let $F(z)$ be the quotient of two polynomials in z satisfying

  (i) $F(z)$ has no poles on the positive real axis, and
  (ii) $z^{a+1}F(z)$ vanishes as z tends to 0 or ∞, where a is real but not an integer.

Then

$$\int_0^\infty x^a F(x)\,dx = \dfrac{2\pi i}{1 - e^{2\pi i a}} \sum_{z \neq 0} \text{Res}(z^a F(z)),$$

the sum being take over all nonzero poles of $F(z)$.

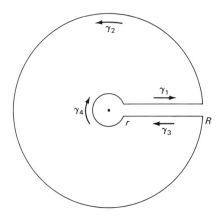

Figure 4.5

Proof Since $F(z)$ has only finitely many poles in $\mathscr{C}$, there are numbers $0 < r < R$ such that all the nonzero poles will be inside the annulus $r < |z| < R$. For the function z^a select the branch of $\mathscr{R}$ whose argument lies between 0 and 2π, with branch points 0 and ∞. Let $\gamma = \gamma_1 + \gamma_2 + \gamma_3 + \gamma_4$ consist of the boundary of the domain obtained by cutting $r < |z| < R$ along the linear segment $r < x < R$, labeled as shown in Figure 4.5.

Strictly speaking, we cannot apply the Residue Theorem directly to γ since $z^a F(z)$ is multivalued on the branch cut. However, we can apply the Residue Theorem to the boundaries of the domains D_1, D_2 indicated in Figure 4.6, with the integrals along the arcs γ_5 and γ_6 canceling out, thus extending the Residue Theorem to γ.

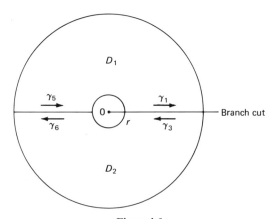

Figure 4.6

4.4 INTEGRATION OF MULTIVALUED FUNCTIONS

Note that the integrand has different values on γ_1 and γ_3. By the Residue Theorem

$$\int_\gamma z^a F(z)\, dz = 2\pi i \sum_{z \neq 0} \text{Res } z^a F(z),$$

but

$$\left| \int_{\gamma_j} z^a F(z)\, dz \right| \leq \int_0^{2\pi} |z^{a+1} F(z)|\, d\theta, \qquad j = 2, 4,$$

which vanishes by (ii), as $R \to \infty$ or $r \to 0$. Now

$$z^a F(z) = \begin{cases} x^a F(x) & \text{on } \gamma_1, \\ x^a e^{2\pi i a} F(x) & \text{on } \gamma_3, \end{cases}$$

so

$$\int_{\gamma_1 + \gamma_3} z^a F(z)\, dz = (1 - e^{2\pi i a}) \int_r^R x^a F(x)\, dx,$$

yields the required formula by letting $R \to \infty$ and $r \to 0$.

Example 1

$$\int_0^\infty \frac{x^a\, dx}{x + b} = \frac{-\pi b^a}{\sin \pi a}, \qquad 0 > a > -1, \quad b > 0.$$

Here $0 < a + 1 < 1$ so it is clear that (i) and (ii) hold. Selecting the branch of $\mathscr{R}$ whose argument lies between 0 and 2π we have

$$\frac{2\pi i}{1 - e^{2\pi i a}} \text{Res}_{-b} \frac{z^a}{(z + b)} = \frac{2\pi i b^a}{e^{-\pi i a} - e^{\pi i a}},$$

since on this branch

$$(-b)^a = b^a e^{\pi i a}.$$

The same type of procedure can be applied to other multivalued functions We illustrate this here and in the next section with several examples:

Example 2

$$\int_0^\infty \frac{\log x}{x^2 + b^2}\, dx = \frac{\pi}{2b} \log b, \qquad b > 0.$$

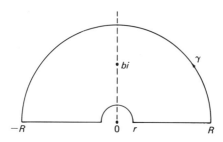

Figure 4.7

Here we use the curve γ shown in Figure 4.7. Then

$$\int_\gamma \frac{\log z}{z^2 + b^2} dz = 2\pi i \operatorname{Res}_{bi} \frac{\log z}{z^2 + b^2} = \frac{\pi}{b}\left[\log b + \frac{i\pi}{2}\right].$$

But

$$\left| iR \int_0^\pi \frac{\log R + i\theta}{(Re^{i\theta})^2 + b^2} e^{i\theta} d\theta \right| \leq \frac{R(\log R + \pi)}{R^2 - b^2} \pi,$$

which vanishes as $R \to \infty$ or 0 by L'Hospital's Rule. Since the integral is convergent

$$\frac{\pi}{b}\left[\log b + \frac{i\pi}{2}\right] = \int_{-\infty}^\infty \frac{\log x \, dx}{x^2 + b^2}$$

$$= \int_{-\infty}^\infty \frac{\log |x|}{x^2 + b^2} dx + i\pi \int_{-\infty}^0 \frac{dx}{x^2 + b^2},$$

from which the result follows as the first integrand is even.

EXERCISES

Evaluate the integrals below using the method of this section.

1. $\int_0^\infty \frac{x^a}{(x+b)^2} dx = \frac{\pi a \, b^{a-1}}{\sin \pi a}$, $\quad 1 > a > -1, \quad b > 0$.

2. $\int_{-\infty}^\infty \frac{e^{ay} \, dy}{1 + be^{-y}} = \frac{-\pi b^a}{\sin \pi a}$, $\quad 0 > a > -1, \quad b > 0$.

3. $\int_0^\infty \frac{x^a \, dx}{x^2 + b^2} = \frac{\pi b^{a-1}}{2 \cos \dfrac{\pi a}{2}}$, $\quad 1 > a > -1, \quad b > 0$.

4.5 OTHER INTEGRATION TECHNIQUES

4. $\displaystyle\int_0^\infty \frac{x^a \, dx}{x^2 + 2x\cos\theta + 1} = \frac{\pi}{\sin \pi a} \frac{\sin \theta a}{\sin \theta}, \qquad 1 > a > -1, \quad \pi > \theta > -\pi.$

5. $\displaystyle\int_0^\infty \frac{x^a \, dx}{(x^2 + b^2)^2} = \frac{\pi b^{a-3}(1-a)}{4\cos\frac{\pi a}{2}}, \qquad 3 > a > -1, \quad b > 0.$

6. $\displaystyle\int_0^\infty \frac{x^a \, dx}{x^3 + b^3} = \frac{2\pi b^{a-2}}{3\sin \pi a}\left[\cos\frac{\pi}{3}(1-2a) - \frac{1}{2}\right], \qquad 2 > a > -1, \quad b > 0.$

7. $\displaystyle\int_0^\infty \frac{\log x}{(x^2 + b^2)^2} \, dx = \frac{\pi}{4b^3}(\log b - 1), \qquad b > 0.$

8. $\displaystyle\int_0^\infty \frac{x^a \log x}{x + b} \, dx = \frac{\pi b^a}{\sin^2 \pi a}(\cos \pi a - \sin \pi a \cdot \log b),$

$\qquad 0 > a > -1, \quad b > 0.$

9. $\displaystyle\int_0^\infty \frac{x^a \log x}{x^2 + b^2} \, dx = \frac{\pi b^{a-1}}{2\cos^2 \frac{\pi a}{2}}\left[\frac{\pi}{2}\sin\frac{\pi a}{2} + \log b \cdot \cos\frac{\pi a}{2}\right],$

$\qquad 1 > a > -1, \quad b > 0.$

10. Prove

$$\int_0^{\pi/2} (\tan\theta)^a \, d\theta = \frac{\pi}{2\cos\frac{\pi a}{2}}, \qquad 1 > a > -1,$$

and

$$\int_0^{\pi/2} \log \tan \theta \, d\theta = 0.$$

(*Hint*: Use Exercise 3.)

4.5† OTHER INTEGRATION TECHNIQUES

Example 1

$$\text{PV} \int_0^\infty \frac{x^a}{x-1} \, dx = -\pi \cot \pi a, \qquad 0 > a > -1.$$

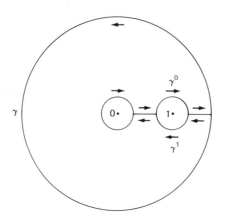

Figure 4.8

The integral of the function $z^a/(z-1)$ vanishes over the curve γ shown in Figure 4.8, as no singularities lie inside γ. The Laurent series of $z^a/(z-1)$ about $e^{2\pi ji}$, $j=0, 1$, depends on the branch of $\mathscr{R}$ selected and is given by

$$\frac{\operatorname{Res}_{e^{2\pi ji}}\left(\dfrac{z^a}{z-1}\right)}{z-e^{2\pi ji}} + f_j(z) = \frac{e^{2\pi jia}}{z-e^{2\pi ji}} + f_j(z),$$

with $f_j(z)$ analytic in $|z - e^{2\pi ji}| < 2\delta$. Using the proof in Section 4.3

$$\int_{\gamma^j} \frac{z^a}{z-1}\, dz = -\pi i e^{2\pi jia} + \int_{\gamma^j} f_j(z)\, dz,$$

with the last integral bounded by $\delta N_j \pi$, where

$$\gamma^j: z = e^{2\pi ji} - (-1)^j \delta e^{-i\theta}, \qquad 0 \le \theta \le \pi, \quad j = 0, 1.$$

Since $z^{a+1}/(z-1)$ vanishes as z tends to 0 or ∞, the integrals along the curves $|z|=R$ and $|z|=r$ vanish as $R \to \infty$ and $r \to 0$. On the line segments z^a takes values that differ by $e^{2\pi ia}$ depending on the branch they belong to. Thus

$$(1 - e^{2\pi ia})\left[\int_0^{1-\delta} + \int_{1+\delta}^{\infty}\right] \frac{x^a}{x-1}\, dx = -\int_{\gamma^0 + \gamma^1} \frac{z^a}{z-1}\, dz,$$

4.5 OTHER INTEGRATION TECHNIQUES

and letting $\delta \to 0$, we have

$$\text{PV} \int_0^\infty \frac{x^a}{x-1} dx = \frac{\pi i}{1 - e^{2\pi i a}} \sum_{j=0}^{1} e^{2\pi j i a}$$

$$= \pi i \frac{1 + e^{2\pi i a}}{1 - e^{2\pi i a}} = -\pi \cot \pi a.$$

This technique can be extended to functions which are quotients of polynomials with poles of order 1 on the positive real axis obtaining the following general result:

Theorem Let $F(z)$ be the quotient of two polynomials in z satisfying

(i) All poles of $F(z)$ on the positive real axis are of order 1, and
(ii) $z^{a+1} F(z)$ vanishes as z tends to 0 or ∞, where a is real but not an integer.

Then

$$\text{PV} \int_0^\infty x^a F(x) \, dx = \frac{2\pi i}{1 - e^{2\pi i a}} \sum_{z \neq x \geq 0} \text{Res } z^a F(z) - \pi \cot \pi a \sum_{z = x > 0} z^a \text{ Res } F(z).$$

Example 2

$$\int_0^\infty \frac{\sinh ax}{\sinh \pi x} dx = \frac{1}{2} \tan \frac{a}{2}, \quad -\pi < a < \pi.$$

The integral of the function $e^{az}/\sinh \pi z$ vanishes over the curve γ shown in Figure 4.9, as no singularities lie inside γ. But

$$|\sinh \pi(R + iy)| \geq |\sinh \pi R|$$

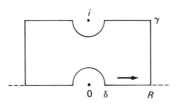

Figure 4.9

(see Exercise 7(e), Section 1.6), implying that

$$\left| \int_R^{R+i} \frac{e^{az}}{\sinh \pi z} dz \right| \le \frac{e^{aR}}{|\sinh \pi R|} \to 0,$$

as $R \to \pm\infty$. Since $i \sinh z = \sin iz$, $1/\sinh \pi z$ has poles of order 1 at all integral multiples of i, thus

$$\text{Res}_0 \frac{e^{az}}{\sinh \pi z} = \lim_{z \to 0} \frac{z e^{az}}{\sinh \pi z} = \frac{1}{\pi},$$

and

$$\text{Res}_i \frac{e^{az}}{\sinh \pi z} = \lim_{z \to i} \frac{(z-i)e^{az}}{\sinh \pi z} = \frac{-e^{ai}}{\pi},$$

by L'Hospital's Rule (Exercise 8, Section 3.2). Integrating over the two semicircles yields

$$-\pi i \left(\frac{1}{\pi} - \frac{e^{ai}}{\pi} \right)$$

plus an integral that vanishes as $\delta \to 0$. But

$$\sinh \pi(x+i) = -\sinh \pi x,$$

so we obtain

$$\text{PV}(1 + e^{ai}) \int_{-\infty}^{\infty} \frac{e^{ax}}{\sinh \pi x} dx = i(1 - e^{ai}),$$

or

$$\text{PV} \int_{-\infty}^{\infty} \frac{e^{ax}}{\sinh \pi x} dx = \tan \frac{a}{2},$$

from which the result follows since the integrand is even.

EXERCISES

Evaluate the following integrals by the methods of this section.

1. $\text{PV} \int_{-\infty}^{\infty} \frac{e^{ax} dx}{e^x - 1} = -\pi \cot \pi a, \quad 1 > a > 0.$

2. $\text{PV} \int_0^{\infty} \frac{x^a dx}{x^2 - b^2} = \frac{\pi b^{a-1}}{2 \sin \pi a} [1 - \cos \pi a], \quad 1 > a > -1, \quad b > 0.$

3. $\text{PV} \int_0^{\infty} \frac{x^a dx}{(x^2 - b^2)(x^2 - c^2)} = \frac{\pi(b^{a-1} - c^{a-1})(1 - \cos \pi a)}{2(b^2 - c^2) \sin \pi a},$
$3 > a > -1, \quad b, c > 0, \quad b \ne c.$

4.6 THE ARGUMENT PRINCIPLE

4. $\int_0^\infty \dfrac{x}{\sinh x}\, dx = \dfrac{\pi^2}{4}.$

5. $\int_0^\infty \dfrac{\sin ax}{\sinh x}\, dx = \dfrac{\pi}{2} \tanh \dfrac{a\pi}{2},$ a real.

6. $\int_0^\infty \dfrac{x \cos ax}{\sinh x}\, dx = \dfrac{\pi^2}{4} \operatorname{sech}^2 \dfrac{a\pi}{2},$ a real.

7. $\int_0^\infty \dfrac{\cosh ax}{\cosh \pi x}\, dx = \dfrac{1}{2} \sec \dfrac{a}{2},$ $-\pi < a < \pi.$

8. $\int_0^\infty x^{a-1} \begin{Bmatrix} \cos bx \\ \sin bx \end{Bmatrix} dx = \begin{Bmatrix} \cos \dfrac{\pi a}{2} \\ \sin \dfrac{\pi a}{2} \end{Bmatrix} \cdot \dfrac{\Gamma(a)}{b^a},$ $1 > a > 0,\ b > 0,$

where $\Gamma(a) = \int_0^\infty e^{-x} x^{a-1}\, dx$ is the *Gamma function*. (*Hint*: Integrate $z^{a-1} e^{-bz}$ around a suitable contour and use the inequality $\cos \theta \geq 1 - (2/\pi)\theta,\ 0 \leq \theta \leq \pi/2$.)

9. $\int_0^\infty \dfrac{\sin x^a}{x^a}\, dx = \dfrac{\Gamma\left(\dfrac{1}{a}\right) \cos \dfrac{\pi}{2a}}{a - 1},$ $1 > a > \dfrac{1}{2}.$

(*Hint*: Show $x\Gamma(x) = \Gamma(x + 1)$ by differentiating by parts.)

10. $\int_0^\infty \dfrac{\cos x}{\sqrt{x}}\, dx = \int_0^\infty \dfrac{\sin x}{\sqrt{x}}\, dx = \int_0^\infty \dfrac{e^{-x}}{\sqrt{2x}}\, dx.$

4.6 THE ARGUMENT PRINCIPLE

Another application of the Residue Theorem, useful in determining the number of zeros and poles of a meromorphic function, is:

The Argument Principle Let $w = f(z)$ be meromorphic in the simply connected domain G and γ be a pwd Jordan curve in G which avoids the zeros and poles of $f(z)$. Then

$$\frac{1}{2\pi i} \int_{f(\gamma)} \frac{dw}{w} = \frac{1}{2\pi i} \int_\gamma \frac{f'(z)\, dz}{f(z)} = Z - P,$$

where Z and P are the number of zeros and poles, including multiplicities, of $f(z)$ lying inside γ, respectively.

Proof Note that the first integral indicates the number of times the closed curve $f(\gamma)$ winds around 0; in other words, it measures the *variation of the argument* of $f(z)$ as z traverses the curve γ, leading to the name of the theorem.

If a is a zero of order k of $f(z)$, then we can write $f(z) = (z-a)^k f_0(z)$, with $f_0(z)$ analytic and nonzero in an ε-neighborhood of a. Thus

$$\frac{f'(z)}{f(z)} = \frac{k}{z-a} + \frac{f_0'(z)}{f_0(z)},$$

and, since f_0'/f_0 is analytic in an ε-neighborhood of a, we see that f'/f has a pole of order 1 with residue k at $z = a$.

On the other hand, if a is a pole of order h of $f(z)$, then $f(z) = f_0(z)/(z-a)^h$ with $f_0(z)$ again nonzero and analytic in an ε-neighborhood of a. So

$$\frac{f'(z)}{f(z)} = \frac{-h}{z-a} + \frac{f_0'(z)}{f_0(z)},$$

has a pole of order 1 with residue $(-h)$ at $z = a$. By the Residue Theorem it follows that

$$\frac{1}{2\pi i} \int_\gamma \frac{f'(z)}{f(z)}\, dz = Z - P,$$

where Z is the sum of all the orders k of the zeros of $f(z)$, and P is the sum of all the orders h of the poles of $f(z)$, lying inside γ.

A most useful application of the Argument Principle is the following result:

Rouché's Theorem Let $f(z)$ and $g(z)$ be analytic in a simply connected domain G. If $|f(z)| > |g(z) - f(z)|$ at all points of the pwd Jordan curve γ lying in G, then $f(z)$ and $g(z)$ have the same number of zeros inside γ.

Proof The hypothesis $|f(z)| > |g(z) - f(z)|$ forces both functions to be nonzero on γ, thus γ avoids the poles and zeros of $F(z) = g(z)/f(z)$. However, for all z on γ

$$\left| \frac{g(z)}{f(z)} - 1 \right| < 1,$$

Thus $F(\gamma)$ does not wind around 0, so the Argument Principle implies that $F(z)$ has the same number of zeros as it has poles inside γ. But these correspond to the zeros of $g(z)$ and $f(z)$ respectively, so the proof is complete.

4.6 THE ARGUMENT PRINCIPLE

Example 1 Find the number of roots of the equation

$$z^4 + 5z + 1 = 0$$

lying inside the circle $|z| = 1$.

Let $f(z) = 5z$ and $g(z) = z^4 + 5z + 1$. Then, by the Triangle Inequality,

$$|g(z) - f(z)| \leq |z|^4 + 1 < |5z| = |f(z)|$$

on $|z| = 1$. Since $f(z)$ has one zero inside $|z| = 1$, so does $g(z)$. On the other hand, letting $f(z) = z^4$, we have

$$|5z + 1| \leq 11 < 16 = |z|^4$$

on $|z| = 2$. Thus $g(z)$ has four zeros inside $|z| = 2$, three of which lie in the annulus $1 < |z| < 2$, since no zeros lie on $|z| = 1$.

Example 2 Show that $z - e^z + a = 0$, $a > 1$, has one root in the left half plane.

Let $f(z) = z + a$ and $g(z) = z - e^z + a$. For $z = iy$ or $|z| = R > 2a$, $x < 0$, we have

$$|g(z) - f(z)| = e^{\operatorname{Re} z} \leq 1 < a < |f(z)|,$$

and $f(z)$ has only one root (at $z = -a$), so the proof is complete.

Example 3 Find the number of roots of the equation

$$z^4 + iz^3 + 3z^2 + 2iz + 2 = 0$$

lying in the upper half plane.

Let $f(z) = z^4 + 3z^2 + 2 = (z^2 + 2)(z^2 + 1)$ and $g(z) = z^4 + iz^3 + 3z^2 + 2iz + 2$. For $z = x$ or $|z| = R \geq 2$ we have

$$|g(z) - f(z)| = |z| \, |z^2 + 2| < |z^2 + 1| \, |z^2 + 2| = |f(z)|,$$

so $g(z)$ has two roots in the upper half plane.

Example 4 Find the number of roots of the equation

$$7z^3 - 5z^2 + 4z - 2 = 0$$

in the disk $|z| \leq 1$.

If we multiply the equation by $z + 1$ we obtain

$$7z^4 + 2z^3 - z^2 + 2z - 2 = 0.$$

Letting $f(z) = 7z^4$ and $g(z) = 7z^4 + 2z^3 - z^2 + 2z - 2$, we find by the Triangle Inequality that

$$|g(z) - f(z)| \le 2|z|^3 + |z|^2 + 2|z| + 2 < 7|z|^4 = |f(z)|,$$

whenever $|z| = 1 + \varepsilon$, $\varepsilon > 0$. Hence $g(z)$ has four roots in $|z| \le 1$, implying the original equation had three roots in the closed unit disk.

EXERCISES

1. Find the number of roots of the following equations inside the circle $|z| = 1$:
   (a) $z^5 + 8z + 10 = 0$.
   (b) $z^8 - 2z^5 + z^3 - 8z^2 + 3 = 0$.
   (c) $z^6 + 3z^5 - 2z^2 + 2z - 9 = 0$.
   (d) $z^7 - 7z^6 + 4z^3 - 1 = 0$.

2. How many of the roots of the equations above lie inside $|z| = 2$?
3. How many roots of the equation

$$3z^4 - 6iz^3 + 7z^2 - 2iz + 2 = 0$$

lie in the upper half plane?

4. How many roots of the equation

$$z^6 + z^5 - 6z^4 - 5z^3 + 10z^2 + 5z - 5 = 0$$

lie in the right half plane?

5. Find the number of roots of the equation

$$9z^4 + 7z^3 + 5z^2 + z + 1 = 0$$

lying in the disk $|z| \le 1$.

6. How many roots of the equation

$$z^4 + 2z^3 - 3z^2 - 3z + 6 = 0$$

lie in the disk $|z| \le 1$?

7. Show the function

$$f(z) = \frac{z - a}{1 - \bar{a}z}, \qquad |a| < 1,$$

takes in $|z| < 1$ every value c satisfying $|c| < 1$ exactly once, and no value c for which $|c| > 1$. Thus $f(z)$ maps $|z| < 1$ one-to-one and onto itself. (*Hint*: Show $|f(z)| = 1$ on $|z| = 1$ and apply Rouché's Theorem to $f(z) - c$.)

NOTES

8. Assuming the hypothesis of the Argument Principle, show that the number of times $f(\gamma)$ winds around the point a equals $Z_a - P$, where Z_a is the number of a values of $f(z)$ including multiplicities.
9. Let $f(z)$ be analytic in a domain G, a in G, and suppose $f(z) - f(a)$ has a zero of order n at $z = a$. Then for sufficiently small $\varepsilon > 0$, there is a $\delta > 0$ such that for all ζ in $|\zeta - f(a)| < \delta$ the equation $f(z) - \zeta = 0$ has exactly n roots in $|z - a| < \varepsilon$.
10. Use the result in Exercise 9 to prove nonconstant analytic functions map open sets onto open sets, and use this fact to get an immediate proof of the Maximum Principle. (*Hint*: Show interior points are mapped onto interior points.)

NOTES

Section 4.1 The results in this section are easily extended to arbitrary closed curves γ in $\mathscr{C}$. However, in most applications γ is a Jordan curve. Consequently this extra hypothesis was incorporated in order to simplify the statements of the theorems. For the more general statements see [A, pp. 147–151].

Sections 4.1–4.5 The reader may have noticed that a number of the integrals, which depended on one or more arbitrary parameters, could be obtained by differentiation or integration of other integrals with respect to these parameters. For example: Exercises 1 and 2 of Section 1 and 2, Exercises 8 and 9 of Section 2, Exercises 5 and 6 of Section 3, Exercises 3, 5, and 9 of Section 4, Exercises 2 and 3 of Section 5. A sufficient condition for the validity of these procedures is the uniform convergence of the integrals on the interval of definition of the parameters. The relevant theorems and proofs may be found in most advanced calculus books. For example see [B, pp. 204–212]. This technique is usually easier than evaluating the integrals by the residue method.

Section 4.6 These results may also be extended to arbitrary closed curves γ (see [A, pp. 151–153]).

Chapter 5 | CONFORMAL MAPPINGS

5.1 GENERAL PROPERTIES

Let us investigate the change in the slope of a differentiable arc passing through the point z_0 under the mapping $w = f(z)$ when f is analytic at z_0 and $f'(z_0) \neq 0$.

If $\gamma: z = z(t)$, $z(0) = z_0$, is such an arc, its tangent at z_0 has slope

$$\frac{dy}{dx} = \frac{y'(0)}{x'(0)} = \tan \arg z'(0),$$

and its image $f(\gamma): w = f(z(t))$ has a tangent at $f(z_0)$ with slope $\tan \arg w'(0)$.

5.1 GENERAL PROPERTIES

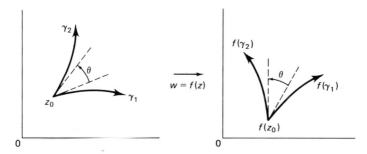

Figure 5.1
Mapping conformal at z_0.

But by the chain rule

$$\arg w'(0) = \arg[f'(z_0)z'(0)] = \arg f'(z_0) + \arg z'(0),$$

indicating the change in direction is always the constant $\arg f'(z_0)$ independent of the arc chosen. Thus, the angle formed by the tangents of two differentiable arcs intersecting at z_0 is preserved under the mapping $w = f(z)$, as both directions are changed by the same amount (see Figure 5.1). Functions that preserve angle size and orientation are said to be *conformal*. We have proved:

Theorem If $f(z)$ is analytic in a domain G, then it is conformal at all points z_0 in G for which $f'(z_0) \neq 0$.

In particular consider the effect of the mapping upon a disk centered at z_0 lying in G (see Figure 5.2). The angles between radial lines are preserved

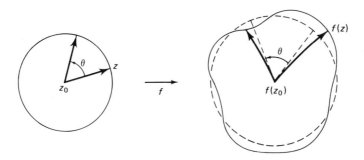

Figure 5.2
Mapping a disk centered at z_0.

although their lengths are not. However, since

$$|f'(z_0)| = \lim_{z \to z_0} \frac{|f(z) - f(z_0)|}{|z - z_0|},$$

the radial lines are subject to approximately the same change of scale $|f'(z_0)|$ when the radius is small. Roughly, *"small"* circles about z_0 *are mapped onto "small" circles about* $f(z_0)$ *with change of scale* $|f'(z_0)|$. Moreover, this indicates that the mapping is *locally* one-to-one, although it is clear nothing can be said about its *global* behavior. For example, $f(z) = z^2$ is one-to-one on any ε-neighborhood of i that remains in the upper half plane, but $f(i) = f(-i)$, so it is not one-to-one in $\mathscr{C}$. Furthermore, circles around the origin are mapped to circular curves which wind around the origin twice. This motivates the following theorem:

Theorem Let $f(z)$ be analytic in a domain G containing the point z_0 at which $f'(z)$ has a zero of order k. Then all angles at z_0 are magnified by the factor $k + 1$.

Proof We can write $f'(z) = (z - z_0)^k g(z)$, with g analytic and nonzero in an ε-neighborhood of z_0. Thus the terms $f'(z_0), f''(z_0), \ldots, f^{(k)}(z_0)$ all vanish in the Taylor series for $f'(z)$. Hence, we have the Taylor series

$$f(z) = f(z_0) + \frac{f^{(k+1)}(z_0)}{(k+1)!}(z - z_0)^{k+1} + \cdots,$$

implying that

$$\arg[f(z) - f(z_0)] = (k+1)\arg(z - z_0) + \arg\left[\frac{f^{(k+1)}(z_0)}{(k+1)!} + \cdots\right].$$

The first two arguments compare the angles between the horizontal direction and the vector pointing from $f(z_0)$ to $f(z)$ and from z_0 to z. If z tends to z_0 along a fixed vector making an angle θ with the horizontal direction, the angle of the vector from $f(z_0)$ to $f(z)$ with the horizontal tends to

$$(k+1)\theta + \arg\left[\frac{f^{(k+1)}(z_0)}{(k+1)!}\right],$$

the last argument being independent of θ. Thus the angle between the tangents of two differentiable arcs intersecting at z_0 is magnified by the factor $k + 1$.

5.2 LINEAR FRACTIONAL TRANSFORMATIONS

EXERCISES

1. Indicate where the following mappings are conformal:

   (a) $w = e^z$, (b) $w = \sin z$,

   (c) $w = \dfrac{1}{z}$, (d) $w = z^2 - z$.

2. Show that the image under the mapping $w = z^2$ of the circle $|z - r| = r$, $r > 0$, is the cardioid with polar equation

   $$\rho = 2r^2(1 + \cos\theta).$$

3. Show that the mapping $w = z + 1/z$ maps circles $|z| = r$ onto ellipses

   $$\frac{x^2}{\left(r + \dfrac{1}{r}\right)^2} + \frac{y^2}{\left(r - \dfrac{1}{r}\right)^2} = 1.$$

4. If $w = f(z)$ is an analytic function, show that its Jacobian satisfies

   $$\frac{\partial(u, v)}{\partial(x, y)} = |f'(z)|^2.$$

5. Let $f(z) = u(x, y) + iv(x, y)$ be conformal and have continuous first partial derivatives u_x, u_y, v_x, v_y in a domain G. Show $f(z)$ is analytic in G. (*Hint:* Show the Cauchy–Riemann equations hold.)

5.2 LINEAR FRACTIONAL TRANSFORMATIONS

A simple, but important, type of conformal mapping is given by the expression

$$w = w(z) = \frac{az + b}{cz + d}, \qquad ad - bc \neq 0,$$

where a, b, c, d are complex constants. Such a mapping is called a *linear fractional transformation*. The condition $ad - bc \neq 0$ prevents its derivative

$$w' = \frac{ad - bc}{(cz + d)^2}$$

from vanishing, as otherwise the function is constant. We can solve for z, obtaining

$$z = \frac{-dw + b}{cw - a},$$

and using the convention that $w(-d/c) = \infty$ and $w(\infty) = a/c$, it follows that w maps $\mathcal{M}$ (the Riemann sphere) one-to-one onto itself. Moreover, the mapping is conformal except at $z = \infty, -d/c$, because at these points either $w'(z)$ or $z'(w)$ vanishes. A composition of two linear fractional transformations is again a linear fractional transformation, since

$$\frac{a\left(\frac{\alpha z + \beta}{\gamma z + \delta}\right) + b}{c\left(\frac{\alpha z + \beta}{\gamma z + \delta}\right) + d} = \frac{(a\alpha + b\gamma)z + (a\beta + b\delta)}{(c\alpha + d\gamma)z + (c\beta + d\delta)}$$

with

$$(a\alpha + b\gamma)(c\beta + d\delta) - (a\beta + b\delta)(c\alpha + d\gamma) = (ad - bc)(\alpha\delta - \beta\gamma) \neq 0.$$

Any linear fractional transformation is a composition of four special types of such transformations:

(i) *Translation*: $w = z + \alpha$, α complex.
(ii) *Rotation*: $w = e^{i\theta}z$, θ real.
(iii) *Magnification*: $w = kz$, $k > 0$.
(iv) *Inversion*: $w = 1/z$.

If $c \neq 0$, we can write

$$\frac{az + b}{cz + d} = \frac{bc - ad}{c^2\left(z + \frac{d}{c}\right)} + \frac{a}{c},$$

showing the transformation may be decomposed into a translation by d/c, followed by a rotation by $e^{2i\arg c}$, a magnification $|c|^2$, an inversion, a rotation, a magnification, and a translation. If $c = 0$,

$$\frac{az + b}{d} = \frac{a}{d}\left(z + \frac{b}{a}\right),$$

proving the decomposition consists of a translation, a rotation, and a magnification.

5.2 LINEAR FRACTIONAL TRANSFORMATIONS

The fundamental property of linear fractional transformations is that *they map circles onto circles* in $\mathcal{M}$. A "circle" in $\mathcal{M}$ corresponds to a circle or a straight line in $\mathscr{C}$, as lines in the plane correspond to circles through ∞ on the Riemann sphere (see Section 1.1). Geometrically it is clear that translations and rotations carry "circles" onto "circles." Before considering the other two transformations, observe that the line $y = \tan\theta \cdot x + b$ can be written in the form

$$\operatorname{Re}(-ie^{i\theta}z) = y\cos\theta - x\sin\theta = b\cos\theta, \qquad |\theta| < \frac{\pi}{2}.$$

The magnification $w = kz$, $k > 0$, maps (by substitution) circles $|z - z_0| = r$ to circles $|w - kz_0| = kr$, and lines $\operatorname{Re}(\alpha z) = c$ to lines $\operatorname{Re}(\alpha w) = ck$, with $|\alpha| = 1$, c real. Under inversion the circle $|z - z_0| = r\ (>0)$ satisfies

$$0 = |z - z_0|^2 - r^2 = |z|^2 + |z_0|^2 - 2\operatorname{Re} z\bar{z}_0 - r^2$$

$$= \frac{1}{|w|^2} + (|z_0|^2 - r^2) - \frac{2}{|w|^2}\operatorname{Re}\bar{z}_0\bar{w}. \qquad (1)$$

If $|z_0| = r$, indicating the circle passes through the origin, we obtain the equation

$$0 = \frac{1 - 2\operatorname{Re}\bar{z}_0\bar{w}}{|w|^2}, \qquad (2)$$

yielding the line $\operatorname{Re}(z_0 w) = \tfrac{1}{2}$ through ∞. If $|z_0| \neq r$, the origin does not lie on the circle, so multiplying Equation (1) by the nonzero quantity $|w|^2/(|z_0|^2 - r^2)$ we have

$$0 = \frac{1}{|z_0|^2 - r^2} + |w|^2 - \frac{2}{|z_0|^2 - r^2}\operatorname{Re}\overline{z_0 w}$$

$$= \left| w - \frac{z_0}{|z_0|^2 - r^2} \right|^2 - \frac{r^2}{(|z_0|^2 - r^2)^2},$$

a circle. That lines map to circles through the origin follows by reversing the steps leading to Equation (2).

Since any linear fractional transformation is a composition of these special transformations we have proved:

Theorem Linear fractional transformations map circles onto circles in $\mathcal{M}$.

5 CONFORMAL MAPPINGS

Example 1 Map the common overlap of the disks $|z - 1| < 1$ and $|z - i| < 1$ conformally onto the first quadrant.

Since the circles $|z - 1| = 1$, $|z - i| = 1$ intersect at the points 0 and $1 + i$, consider the mapping

$$\zeta = \frac{z}{z - (1 + i)},$$

which sends 0 to 0 and $1 + i$ to ∞. The circles map to lines perpendicular to each other at the origin, since the mapping is conformal and the tangent lines to the circles are perpendicular at $z = 0$. Since $\zeta(2) = 1 + i$ and $\zeta((1 + i)/2) = -1$, the lines have slope ± 1 in the ζ-plane and the overlap corresponds to the set $|\arg \zeta - \pi| < \pi/4$ (see Figure 5.3). The rotation

$$w = e^{-3\pi i/4}\zeta = \frac{e^{-3\pi i/4}z}{z - (1 + i)}$$

yields the desired mapping.

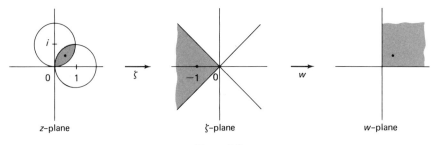

z-plane   ζ-plane   w-plane

Figure 5.3

Example 2 Map the right half plane onto the unit disk $|z| < 1$ so that the point 1 is mapped to the origin.

Observe that the mapping

$$w = \frac{z - 1}{z + 1} \tag{3}$$

sends 1 to 0, 0 to -1, and ∞ to 1. Moreover, since -1 is mapped to ∞, the imaginary axis is mapped to a circle in $\mathscr{C}$ passing through the points ± 1. On the other hand, the real axis is mapped onto itself. Since the real and imaginary axes are perpendicular to each other at the origin, the tangent line to the circle at $w = -1$ is vertical, implying the circle is in fact the unit circle $|w| = 1$.

5.2 LINEAR FRACTIONAL TRANSFORMATIONS

Example 3 Find the number of roots of the equation

$$p(z) = 11z^4 - 10z^3 - 4z^2 + 10z + 9 = 0$$

lying in the right half plane.

Since transformation (3) maps the right half plane onto the disk, by substituting the inverse mapping

$$z = \frac{1+w}{1-w},$$

we obtain the equivalent problem of finding the number of roots of the equation

$$p(z(w)) = 16(w^4 + 3w^3 + 8w^2 - 2w + 1) = 0$$

lying in $|w| < 1$. Letting $f(w) = 8w^2$ and $g(w) = p(z(w))/16$ we find that

$$|g(w) - f(w)| \leq 7 < 8|w|^2 = |f(w)|$$

on $|w| = 1$, implying by Rouché's Theorem that $p(z)$ has two roots in the right half plane.

EXERCISES

In Exercises 1–4 describe the image of the domain indicated under the given mapping.

1. The disk $|z| < 1$; $w = i\dfrac{z-1}{z+1}$.

2. The quadrant $x > 0$, $y > 0$; $w = \dfrac{z-i}{z+i}$.

3. The angular sector $|\arg z| < \dfrac{\pi}{4}$; $w = \dfrac{z}{z-1}$.

4. The strip $0 < x < 1$; $w = \dfrac{z}{z-1}$.

5. Find the number of roots of the equation

$$11z^4 - 20z^3 + 6z^2 + 20z - 1 = 0$$

lying in the right half plane.

6. How many roots of the equation

$$17z^4 + 26z^3 + 56z^2 + 38z + 7 = 0$$

lie in the first quadrant?

7. Using the exponential function, map the domain lying inside $|z| = 2$ and outside $|z - 1| = 1$ onto the upper half plane.
8. Map the domain $|z - 1| < 1$, $\text{Im } z < 0$, onto the upper half plane.
9. Map the sector $|\arg z| < \pi/4$ onto the set $|\text{Re } w| < 1$, $\text{Im } w > 0$. (*Hint*: Use the sine function.)

5.3 CONTINUATION

Given three distinct points z_1, z_2, z_3 in $\mathcal{M}$, there is a linear fractional transformation carrying them into 0, 1, ∞, respectively. If none of the points is ∞, it is given by

$$w = \frac{(z - z_1)(z_2 - z_3)}{(z - z_3)(z_2 - z_1)},$$

and becomes

$$\frac{z_2 - z_3}{z - z_3}, \qquad \frac{z - z_1}{z - z_3}, \qquad \frac{z - z_1}{z_2 - z_1},$$

if z_1, z_2, or $z_3 = \infty$. If w^* is another linear fractional transformation with the same property, then the composition $w^* w^{-1}$ keeps the points 0, 1, ∞ fixed. Thus we have a linear fractional transformation

$$\zeta = \frac{az + b}{cz + d}, \qquad ad - bc \neq 0,$$

satisfying the equations

$$0 = \frac{b}{d}, \qquad 1 = \frac{a + b}{c + d}, \qquad \infty = \frac{a}{c}.$$

But then $b = c = 0$ and $a = d$ implying $w^* w^{-1} = I$, the identity mapping, and hence $w^* = w$. Therefore w *is the only linear fractional transformation mapping the points* z_1, z_2, z_3 *into* 0, 1, ∞, *respectively*.

Since a circle is determined by three of its points, we can now easily determine a linear fractional transformation carrying a given circle in the z-plane into a given circle in the w-plane. Select distinct points z_1, z_2, z_3 on the first circle and w_1, w_2, w_3 on the second circle, then

$$\frac{(w - w_1)(w_2 - w_3)}{(w - w_3)(w_2 - w_1)} = \frac{(z - z_1)(z_2 - z_3)}{(z - z_3)(z_2 - z_1)}.$$

5.3 CONTINUATION

maps z_1, z_2, z_3 into w_1, w_2, w_3, as the right-hand side of the equation maps z_1, z_2, z_3 into $0, 1, \infty$, and the inverse of the left-hand side maps $0, 1, \infty$ into w_1, w_2, w_3.

The points w and $\bar{w}$ are symmetric with respect to the real axis. We generalize this concept to any circle C in $\mathcal{M}$:

Definition The points z and z^* are *symmetric with respect to the circle C*, in the extended z-plane, if there is a linear fractional transformation w mapping C onto the real axis and satisfying $\overline{w(z)} = w(z^*)$.

At first glance it might appear that symmetry with respect to C depends on the transformation w, but if w^* is a linear fractional transformation also mapping C onto the real axis, then $\zeta = w^*w^{-1}$ maps the real axis onto itself. Therefore, it is of the form

$$\frac{(\zeta - b_1)(b_2 - b_3)}{(\zeta - b_3)(b_2 - b_1)} = \frac{(w - a_1)(a_2 - a_3)}{(w - a_3)(a_2 - a_1)},$$

with $a_j, b_j, j = 1, 2, 3$ real. Solving for ζ we obtain

$$\zeta = \frac{\alpha w + \beta}{\gamma w + \delta},$$

with $\alpha, \beta, \gamma, \delta$ real, thus

$$\overline{w^*(z)} = \overline{\zeta(w(z))} = \zeta(\overline{w(z)}) = \zeta(w(z^*)) = w^*(z^*),$$

and symmetry is independent of the transformation employed. Moreover, *symmetry is preserved under linear fractional transformations*, for if z and z^* are symmetric with respect to the circle C and w^* is any such transformation, then $w^*(z)$ and $w^*(z^*)$ are symmetric with respect to $w^*(C)$ under the mapping ww^{*-1}. This fact is called the *Symmetry Principle*. Finally, it would be useful to have a formula for the point symmetric to z with respect to the circle C centered at a with radius R. The mapping

$$w = i\frac{z - (a - R)}{z - (a + R)}$$

maps C to the real line, thus

$$\overline{\left[i\frac{z - (a - R)}{z - (a + R)}\right]} = i\frac{\frac{R}{\bar{z} - \bar{a}} + 1}{\frac{R}{\bar{z} - \bar{a}} - 1} = i\frac{\left[\frac{R^2}{\bar{z} - \bar{a}} + a\right] - (a - R)}{\left[\frac{R^2}{\bar{z} - \bar{a}} + a\right] - (a + R)},$$

implying $z^* = R^2/(\bar{z} - \bar{a}) + a$, or that
$$(z^* - a)(\bar{z} - \bar{a}) = R^2.$$

This equation yields the identities
$$\arg\left(\frac{z^* - a}{z - a}\right) = \arg\left(\frac{R^2}{|z - a|^2}\right) = 0, \quad \frac{|z - a|}{R} = \frac{R}{|z^* - a|},$$

indicating z and z^* lie on the same ray from a and that the geometric construction of Figure 5.4 holds by trigonometry.

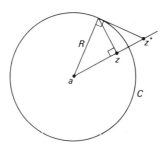

Figure 5.4

Geometric construction of symmetric points with respect to the circle $|z - a| = R$.

Example 1 To find the linear fractional transformation mapping the points $1, i, -1$ into the points $2, 3, 4$, respectively, solve
$$\frac{(w - 2)(3 - 4)}{(w - 4)(3 - 2)} = \frac{(z - 1)(i + 1)}{(z + 1)(i - 1)}$$
for w, obtaining
$$w = \frac{(2 - 4i)z + (2 + 4i)}{(1 - i)z + (1 + i)}.$$

Example 2 The point symmetrical to the point $1 + 2i$ with respect to the circle $|z + 1| = 1$ is obtained by setting $z = 1 + 2i$, $a = -1$, and $R = 1$ in the equation
$$z^* = \frac{R^2}{(\bar{z} - \bar{a})} + a,$$
yielding $z^* = (-3 + i)/4$.

5.3 CONTINUATION

Example 3 Find the linear fractional transformation and the number $a < 1$ mapping the right half plane with the disk $|z - 1| \leq a$ deleted, onto the ring $1 < |w| < 2$.

We begin by finding points z_0, z_0^* symmetric with respect to both the imaginary axis and the circle $|z - 1| = a$. In particular $z_0^* = -z_0$, and since the points $1, z_0, z_0^*$ must lie on a straight line, z_0 is real. Assume that $z_0 > 0$, then by symmetry

$$(-z_0 - 1)(z_0 - 1) = a^2,$$

implying that $z_0 = \sqrt{1 - a^2}$. By the Symmetry Principle, the mapping

$$\zeta = \frac{z - z_0}{z + z_0},$$

sends z_0 to 0 and $-z_0$ to ∞ and maps the imaginary axis and the circle $|z - 1| = a$ onto concentric circles centered at the origin. Since

$$\zeta(\infty) = 1, \qquad \zeta(1 + a) = \frac{1 - z_0}{a} < 1,$$

we magnify by $a/(1 - z_0) = 2$, yielding $a = \frac{4}{5}$ and

$$w = 2\zeta = 2\left(\frac{z - \frac{3}{5}}{z + \frac{3}{5}}\right).$$

EXERCISES

1. Find the linear fractional transformation mapping the points $-1, i, 1 + i$, respectively, into the points:

   (a) $0, 1, \infty$; (b) $1, \infty, 0$; (c) $2, 3, 4$.

2. Is $w = \bar{z}$ a linear fractional transformation?

3. Show any four distinct points can be mapped by a linear fractional transformation to the points $1, -1, k, -k$, where k depends on the original points.

4. Find the points symmetrical to the point $3 + 4i$ with respect to the circles:

   (a) $|z| = 1$; (b) $|z - 1| = 1$; (c) $|z - i| = 2$.

5. Map the unit circle onto itself such that the point α goes into 0, and $\alpha/|\alpha|$ into 1, $|\alpha| < 1$. (*Hint:* Map α^* to ∞.)

6. Find the linear fractional transformation which carries $|z| = 1$ into $|z - 1| = 1$, the point -1 into 0, and 0 into $2i$.

7. Find a linear fractional transformation which carries $|z| = 1$ and $|z - 1| = 3$ into concentric circles. What is the ratio of the radii?
8. Do Exercise 7 for $|z| = 1$ and Im $z = 2$.

5.4 THE SCHWARZ–CHRISTOFFEL FORMULA

The techniques indicated in the previous sections for conformally mapping a given domain onto another leads to the question under what conditions this can be accomplished. The next result, whose proof is beyond the scope of this book, is the fundamental result in this direction:

Riemann Mapping Theorem Let z_0 be a point in a simply connected domain G ($\neq \mathscr{C}$). Then there is a unique analytic function $w = f(z)$ mapping G one-to-one onto the disk $|w| < 1$ such that $f(z_0) = 0$ and $f'(z_0) > 0$.

Now suppose G and G^* are two simply connected domains different from $\mathscr{C}$. The theorem yields the existence of analytic functions f, f^* mapping G, G^* onto the unit disk. Thus $f^{*-1}f$ is a one-to-one mapping of G onto G^*. If we can show f^{*-1}, and thus the composition, is analytic, we then have a conformal mapping of G onto G^*, proving that *any two simply connected domains different from the plane can be mapped conformally onto each other*. Since f^* is conformal (it is one-to-one and analytic), so is f^{*-1}, and the Inverse Function Theorem of calculus (see [B, p. 278]) shows f^{*-1} has continuous first partial derivatives, hence f^{*-1} is analytic (Exercise 5, Section 5.1).

Although the Riemann Mapping Theorem asserts the existence of a function mapping a given domain conformally onto a disk, it does not show how to find it. Construction of the function can be a matter of great difficulty. However, it is possible to obtain an explicit formula for mapping the unit disk onto a polygon. Equivalently, after a linear fractional transformation (see Example 2, Section 5.2), there is an explicit formula for mapping the upper half plane onto a polygon.

Schwarz–Christoffel Formula All functions $w = f(z)$ mapping the upper half plane conformally onto a polygon with exterior angles $\pi\alpha_k$, $k = 1, \ldots, n$, are of the form

$$(z) = A + B \int_0^z \frac{dz}{(z - x_1)^{\alpha_1} \cdots (z - x_n)^{\alpha_n}},$$

5.4 THE SCHWARZ–CHRISTOFFEL FORMULA

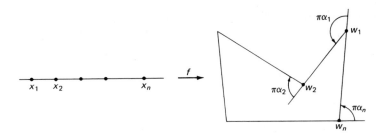

Figure 5.5

Exterior angle of a polygon

where the points $x_1 < x_2 < \cdots < x_n$ lie on the real axis and A, B are complex constants. If $x_n = \infty$, the term involving x_n is left out of the formula.

The *exterior angle* at a vertex of a polygon is the difference between π and the interior angle, in radian measure. As such, $0 < |\alpha_k| < 1$ and $\sum_1^n \alpha_k = 2$ (see Figure 5.5). Note that $\pi\alpha_k$ is the rotation required to bring the direction of the vector from w_{k-1} to w_k into coincidence with the direction of the vector from w_k to w_{k+1}.

The constants A and B control by translation, magnification, and rotation the location, scale, and orientation of the polygon in the w-plane, and the points x_k map to the vertices w_k of the polygon.

Observe that a linear fractional transformation of the upper half plane onto itself allows us to map three of the points x_k onto three prescribed positions on the real axis. Thus, *we are free to select the location of three of the points x_k*. Depending on the polygon, an appropriate choice of the locations of these three points can be extremely useful in obtaining a closed form solution for the integral. The location of the remaining points x_k depends on the shape of the polygon, and can be very difficult to establish except in cases where the polygon is highly symmetric.

The proof of this theorem is complicated, so we will content ourselves by illustrating some examples of its use:

Example 1 Map the upper half plane onto a triangle with exterior angles $\pi\alpha$, $\pi\beta$, $\pi\gamma$, $\alpha + \beta + \gamma = 2$.

Selecting $0, 1, \infty$ as the points we wish to map to the vertices with exterior angles $\pi\alpha$, $\pi\beta$, $\pi\gamma$, respectively, we find the function has the form

$$f(z) = A + B \int_0^z \frac{dz}{z^\alpha(z-1)^\beta}.$$

Since A and B merely affect the position and size of the triangle, in order to find the simplest formula for the location of the vertices set $A = 0$, $B = e^{i\pi\beta}$, and

$$f(z) = \int_0^z \frac{dz}{z^\alpha (1-z)^\beta}.$$

Then $f(0) = 0$ and

$$f(1) = \int_0^1 \frac{dx}{x^\alpha (1-x)^\beta} = \frac{\Gamma(1-\alpha)\Gamma(1-\beta)}{\Gamma(\gamma)}.$$

Since the Gamma function satisfies the identity $\Gamma(x)\Gamma(1-x) = \pi(\sin \pi x)^{-1}$ (see Exercise 2, Section 8.3) the length of this side is

$$c = \frac{1}{\pi} \sin \pi\gamma \; \Gamma(1-\alpha)\Gamma(1-\beta)\Gamma(1-\gamma).$$

Using the law of sines we find the lengths of the other two sides are

$$a = \frac{1}{\pi} \sin \pi\alpha \; \Gamma(1-\alpha)\Gamma(1-\beta)\Gamma(1-\gamma),$$

$$b = \frac{1}{\pi} \sin \pi\beta \; \Gamma(1-\alpha)\Gamma(1-\beta)\Gamma(1-\gamma)$$

(see Figure 5.6).

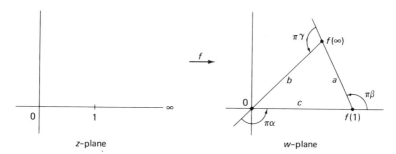

Figure 5.6

Example 2 Map the upper half plane onto a rectangle.

By Exercise 3, Section 5.3, any four points on the real axis may be mapped by a linear fractional transformation onto the points ± 1, $\pm k$, $k > 1$ (invert if necessary). Therefore, such a mapping is given by

$$(z) = \int_0^z \frac{dz}{\sqrt{(1-z^2)(k^2-z^2)}}$$

5.4 THE SCHWARZ–CHRISTOFFEL FORMULA

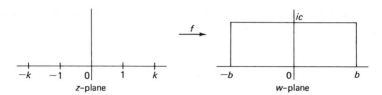

Figure 5.7

(see Figure 5.7). From the formula it is clear the vertices of the rectangle are symmetric with respect to the imaginary axis with

$$b = \frac{1}{k}\int_0^1 \frac{dx}{\sqrt{(1-x^2)(1-k^{-2}x^2)}} = \frac{1}{k} F\left(\frac{\pi}{2}, \frac{1}{k}\right)$$

(an elliptic integral of the first kind), and

$$ic = \int_1^k \frac{dx}{\sqrt{(1-x^2)(k^2-x^2)}} = \frac{i}{k}\int_1^k \frac{dx}{\sqrt{(x^2-1)(1-k^{-2}x^2)}}.$$

We may also apply the Schwarz–Christoffel Formula to map the upper half plane onto a degenerate polygon (that is, having one or more vertices at ∞). In doing so, the exterior angles $\pi\alpha_k$ may assume values outside the range previously indicated. The exterior angle at a vertex at ∞ may be obtained by using the equation $\sum_1^n \alpha_k = 2$ or, in the event that more than one vertex lies at ∞, by examining the polygon in the Riemann sphere $\mathcal{M}$.

Example 3 Find the mapping carrying the upper half plane onto the half strip $|x| < \pi/2$, $y > 0$.

Let the vertices w_1, w_2, w_3 lie at the points $-\pi/2$, $\pi/2$, ∞, respectively (see Figure 5.8). The exterior angles $\pi\alpha_1$ and $\pi\alpha_2$ are both $\pi/2$, so $\alpha_3 = 1$.

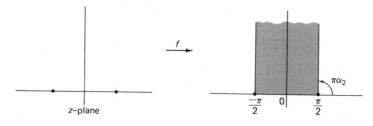

Figure 5.8

Selecting the points x_1, x_2, x_3 to be $-1, +1, \infty$, respectively, we find the transformation is of the form

$$w = f(z) = A + B \int_0^z \frac{dz}{\sqrt{z+1}\sqrt{z-1}}$$

$$= A + iB \int_0^z \frac{dz}{\sqrt{1-z^2}} = A + iB \sin^{-1} z.$$

But the equations

$$-\frac{\pi}{2} = f(-1) = A - i\frac{B\pi}{2}, \quad \frac{\pi}{2} = f(1) = A + i\frac{B\pi}{2}$$

require that $A = 0$ and $B = -i$. Thus $w = \sin^{-1} z$.

EXERCISES

1. Why is it impossible for $G = \mathscr{C}$ in the Riemann Mapping Theorem?
2. Prove every conformal mapping of a disk onto another is given by a linear fractional transformation. Why does this imply the uniqueness of the function in the Riemann Mapping Theorem? (*Hint:* Use Schwarz's Lemma, Exercise 3, Section 2.5.)
3. Show the function

$$f(z) = \int_0^z \frac{dz}{\sqrt{z(z^2-1)}}$$

   maps the upper half plane onto a square with sides of length

$$a = \frac{1}{2} \int_0^1 \frac{dt}{t^{3/4}\sqrt{1-t}} = \frac{\Gamma(\tfrac{1}{4})\Gamma(\tfrac{1}{2})}{2\Gamma(\tfrac{3}{4})} = \frac{\Gamma^2(\tfrac{1}{4})}{2\sqrt{2\pi}}.$$

4. Using the Schwarz–Christoffel transformation, find the mapping carrying the upper half plane onto the infinite strip $|y| < 1$.
5. Map the upper half plane conformally onto the exterior of the half strip $|x| < \pi/2, y > 0$.
6. Map the upper half plane conformally onto the domains indicated in Figure 5.9 with $0, 1, \infty \to \alpha, \beta, \gamma$, respectively.

5.4 THE SCHWARZ–CHRISTOFFEL FORMULA

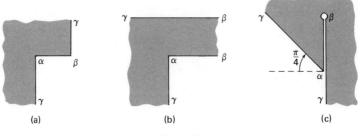

Figure 5.9

7. Map the upper half plane conformally onto the region pictured in Figure 5.10 and show the length (with $|B| = 1$) of the segment from α to β equals $12\pi\sqrt{2\pi}/5\Gamma^2(\frac{1}{4})$.

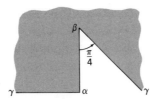

Figure 5.10

8. Map the upper half plane conformally onto the domain in Figure 5.11 with $0, x, 1, \infty \to \alpha, \beta, \gamma, \delta$, respectively, and show $x = k^2$, $0 < k < 1$. [*Hint:* Let $s^2 = (z-1)/(z-x)$.]

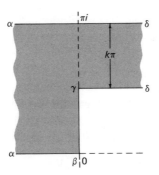

Figure 5.11

9. Show

$$f(z) = A\left[\operatorname{Log}\frac{\sqrt{z}+\sqrt{z-a}}{\sqrt{z}-\sqrt{z-a}} - i\sqrt{a-1}\operatorname{Log}\frac{i\sqrt{z(a-1)}+\sqrt{z-a}}{i\sqrt{z(a-1)}-\sqrt{z-a}}\right] + B$$

maps the upper half plane conformally onto the domain in Figure 5.12 with $0, 1, a, \infty \to \alpha, \beta, \gamma, \delta$ and $a = 1 + (h^2/H^2)$.

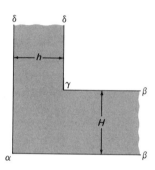

Figure 5.12

5.5 PHYSICAL APPLICATIONS

In this section we will briefly mention some physical problems which can be analyzed using analytic functions.

Since a complex function may be decomposed into two independent real functions, the theory of analytic functions is very useful in solving problems involving two independent variables in two-dimensional space. As this is not a treatise on mathematical physics, much of what follows is a heuristic outline of the physical theory.

Fluid Flow A complete description of the motion of a fluid requires the knowledge of the velocity vector at all points of the fluid at any given time. Suppose the fluid is *incompressible* (that is, of constant density) and the flow is *steady* (independent of time) and *two-dimensional* (the same in all planes parallel to the *xy*-plane in three-dimensional space). Conditions of this type occur, for instance, when the fluid flows past a long cylindrical object whose axis is perpendicular to the direction of the flow. The *velocity* vector can then be given as a continuous complex-valued function of a complex variable $V = V(z)$ for all z in a domain G. We also assume in this section that no *sources* or *sinks* (points at which fluid is being created or destroyed) lie in the domain G.

5.5 PHYSICAL APPLICATIONS

The assumptions that the fluid is incompressible and there are no sources or sinks in G imply that a simply connected domain in G always contains the same amount of fluid. Thus, the quantity of fluid per unit time passing a length element ds on a pwd Jordan curve γ, lying together with its inside in G, is $V_n\, ds$, where V_n is the component (a real number) of V in the outward normal direction to the curve (see Figure 5.13). Hence the total outward *flow*

$$Q = \int_\gamma V_n\, ds \qquad (1)$$

vanishes.

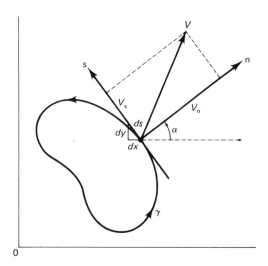

Figure 5.13
Components of the velocity vector.

The line integral of the tangential component V_s of the velocity V around the curve γ,

$$\Gamma = \int_\gamma V_s\, ds, \qquad (2)$$

is called the *circulation* of V along γ. If the circulation is not zero on some curve γ, then the tangential components having one sign dominate the ones having the other sign, in the integral (2). Roughly, this means that the fluid whirls around γ. The flow is said to be *irrotational* if the circulation vanishes along all closed curves in G. We assume the flow is irrotational.

Consider Figure 5.13, where the outward normal and tangential directions to the curve γ are indicated at a point z. Let $\alpha = \alpha(z)$ be the angle between the (positive) horizontal direction and the outward normal to γ at z, and suppose the velocity vector V at z is as indicated.

Rotating the ns-coordinate system about the point z through an angle of $-\alpha$ yields the normal and tangential components of the velocity vector V

$$V_n = \text{Re}(e^{-i\alpha}V), \qquad V_s = \text{Im}(e^{-i\alpha}V).$$

In particular, we have

$$\overline{e^{-i\alpha}V} = V_n + iV_s. \tag{3}$$

The length element ds is related (see Figure 5.13) to the element dx and dy by the identities

$$dx = \cos\left(\frac{\pi}{2} + \alpha\right)ds, \qquad dy = \sin\left(\frac{\pi}{2} + \alpha\right)ds,$$

implying that

$$dz = dx + i\,dy = e^{i(\pi/2 + \alpha)}\,ds = ie^{i\alpha}\,ds. \tag{4}$$

Now, if γ is any pwd Jordan curve contained together with its inside in G, we have by (1)–(4)

$$\int_\gamma \overline{V(z)}\,dz = i\int_\gamma \overline{(e^{-i\alpha}V)}\,ds$$

$$= i\int_\gamma \overline{(V_n + iV_s)}\,ds$$

$$= \int_\gamma (V_s + iV_n)\,ds = 0,$$

implying $\overline{V(z)}$ is analytic by Morera's Theorem. If G is simply connected, the anti-derivative of $\overline{V(z)}$ is an analytic function $w(z) = u(z) + iv(z)$, called the *complex potential* of the flow; u is known as the *potential* function, and v as the *stream* function.[1] Individual particles of the fluid move along curves whose direction at each point coincides with that of the velocity vector. Such curves are called *stream lines* and are characterized by the equation $v(z) = \text{constant}$, since the tangent to such a curve has slope

$$\frac{dy}{dx} = -\frac{v_x}{v_y} = -\frac{v_x}{u_x} = -\tan\arg w' = \tan\arg V,$$

by the Cauchy–Riemann equations.

---

[1] We avoid using the symbol ϕ for the potential function in order to emphasize the analogy between fluid flow and heat flow, and to preserve our mapping notation.

5.5 PHYSICAL APPLICATIONS

The curves $u(z) = $ constant, are called *equipotential lines* and are normal to the streamlines, since

$$\frac{dy}{dx} = -\frac{u_x}{u_y} = \frac{u_x}{v_x} = \frac{-1}{\tan \arg V}.$$

Points at which $V(z) = 0$, and consequently $w'(z) = 0$, are known as *stagnation points* of the flow.

A fundamental problem in fluid dynamics is to construct the flow of an incompressible fluid in a domain G containing no sources or sinks so that the flow has one or more given curves C as stream lines.

Example 1 Suppose we have a uniform flow of velocity A (>0) in the positive x-direction in the upper half plane. This approximates fluid flow in extremely wide channels (see Figure 5.14).

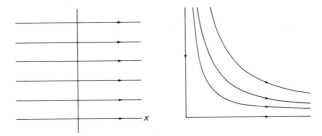

Figure 5.14

Since $V(z) = A$, it follows that $w'(z) = A$, so the complex potential is $w(z) = Az + c$, where $c = c_1 + ic_2$ is a complex constant. Thus, $u(z) = Ax + c_1$ and $v(z) = Ay + c_2$, so the equipotential lines are vertical and the stream lines are horizontal (neglecting the effect of viscosity on the real axis). Setting $c = 0$ the streamline $v = 0$ coincides with the real axis.

If we are interested in finding the stream lines along a right angle in a wide channel, we can approximate this situation by studying the flow in the first quadrant. Observe the mapping $\zeta = z^2$ maps the quadrant onto the upper half plane. Thus, if we know the complex potential $w = w(\zeta)$ of the flow in the upper half plane, $w = w(z^2)$ is the complex potential of the flow in the first quadrant. Assuming we have a uniform flow, as above, in the upper half plane (and $c = 0$) the complex potential in the quadrant satisfies $w = Az^2$, the stream lines are given by the hyperbolas $2Axy = $ constant, and the velocity vector is $V(z) = 2A\bar{z}$.

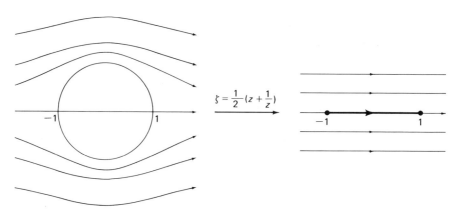

Figure 5.15

Similarly, to find the flow outside the unit disk, notice that $\zeta = (z + z^{-1})/2$ maps this domain onto the complement of the line segment from -1 to 1 (see Figure 5.15). Assuming this segment is part of a streamline, the complex potential is given by

$$w(z) = \frac{A}{2}\left(z + \frac{1}{z}\right).$$

Letting $z = re^{i\theta}$, we find that the stream function is

$$v = \frac{A}{2}\left(r - \frac{1}{r}\right)\sin\theta,$$

the stream line $v = 0$ consists of the unit circle and the real axis with $|x| \geq 1$, and stagnation points occur at ± 1.

Thus, the general procedure is to map the domain G onto one where the flow is known and transforms this flow back to the domain G.

Heat Flow The study of heat conduction in a solid homogenous body may be analyzed in the same manner as was done in the case of fluid flow, if the solid is such that the flow is two-dimensional and the flow of heat is in steady state. Assume no heat sources or sinks are present in the simply connected domain G. Since two points may have different temperatures, there is a flow of heat, by conduction, from the hotter to the cooler parts.

5.5 PHYSICAL APPLICATIONS

The *heat flow* vector $Q = Q(z)$ may be written as a continuous complex-valued function. The heat flowing out of the inside of the pwd closed curve γ contained in G must satisfy

$$\int_\gamma Q_n \, ds = 0,$$

as otherwise the temperature inside will change. Since heat flows from hotter to cooler parts, it is irrotational, so we have

$$\int_\gamma Q_s \, ds = 0,$$

thus $\bar{Q}$ is again analytic in G. Then

$$Q(z) = -k\overline{w'(z)},$$

where k is the *coefficient of thermal conductivity* and $w = u + iv$ is the *complex potential* of the thermal field. By the Cauchy–Riemann equations $\overline{w'(z)} = u_x + iu_y = \text{grad } u(z)$, the *gradient* of u, so $Q = -k \text{ grad } u$ (Fourier's Law), implying the heat flow is normal to the curves $u(z) = \text{constant}$. Points on these lines must then have equal temperature, thus the curves $u(z) = \text{constant}$ are the *isotherms* and $u(z)$ is the *temperature*. The curves $v(z) = \text{constant}$ are called the *stream lines* and are orthogonal to the isotherms.

A frequent problem in steady state heat flow is to construct the isotherms in a domain G with given boundary temperatures.

Example 2 Find the isotherms of the plate G indicated in Figure 5.16 insulated along the segment $0 < x < 1$, $y = 0$ with temperature $0°$ along $z = y \geq 0$, and $1°$ along $z = x \geq 1$.

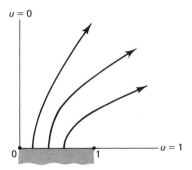

Figure 5.16

The function $w = (2/\pi)\sin^{-1} z$ maps G onto the strip $0 < u < 1$, $v > 0$, and is thus the complex potential. Then

$$z = \sin\frac{\pi}{2}w = \sin\frac{\pi}{2}u \cosh\frac{\pi}{2}v + i\cos\frac{\pi}{2}u \sinh\frac{\pi}{2}v,$$

hence we find

$$\frac{x^2}{\sin^2\frac{\pi}{2}u} - \frac{y^2}{\cos^2\frac{\pi}{2}u} = 1,$$

implying the isotherms are hyperbolas.

Example 3 Find the isotherms of a plate G shaped as in Figure 5.9(a), insulated along the segment joining $\alpha = 0$ to $\beta = 1$, with temperature $-1°$ on the ray from α to γ and $1°$ on the ray from β to γ.

Since the exterior angles at 0 and 1 are $-\pi/2$ and $\pi/2$, respectively, the Schwarz–Christoffel tranformation

$$z = 1 + \frac{i}{\pi}\int_1^\zeta \sqrt{\frac{\zeta+1}{\zeta-1}}\,d\zeta = 1 + \frac{i}{\pi}[\sqrt{\zeta^2-1} + \cosh^{-1}\zeta]$$

maps the upper half plane onto G with $-1, 1, \infty \to \alpha, \beta, \gamma$. But $\zeta = \sin(\pi w/2)$ maps the strip $|u| < 1$, $v > 0$ onto the upper half plane, so

$$z = 1 + \pi^{-1}\left[i\cosh^{-1}\left(\sin\frac{\pi w}{2}\right) - \cos\frac{\pi w}{2}\right]$$

maps the strip above onto G, hence its inverse $w = w(z)$ is the complex potential. As in Example 2, the isotherms will be the images under $z = z(w)$ of the vertical lines $u = $ constant. Simplifying the first term in the parenthesis, we find

$$z = \frac{w+1}{2} - \frac{1}{\pi}\cos\frac{\pi w}{2},$$

from which the isotherms may easily be graphed.

Electrostatics Consider a *plane* electrostatic field $E(z)$ arising from the attraction or repulsion of an arbitrary system of charges (sources and sinks)

5.5 PHYSICAL APPLICATIONS

in the plane. In a simply connected domain G complementary to these charges, the inside of a pwd closed curve γ in G has no charge so

$$\int_\gamma E_n \, ds = 0,$$

by Gauss's Law. The *circulation* of the field is the work done by the field when a positive unit charge is taken completely around the curve γ. As no expenditure of energy is required to maintain an electrostatic field, we have

$$\int_\gamma E_s \, ds = 0.$$

Then E is said to be a *potential field*, $\bar{E}/i$ is analytic and its antiderivative $iw = -v + iu$ is called the *complex potential* of the field; $-v$ is the *force* function and u the *potential* function. By the Cauchy–Riemann equations we find

$$E = -\overline{w'(z)} = -(u_x + iu_y) = -\operatorname{grad} u.$$

The curves $v(z) = \text{constant}$ are the *lines of force*, and $u(z) = \text{constant}$ are *equipotential lines*.

Frequently we wish to find the equipotential lines of a plane electrostatic field bounded by contours on which the potential is a given constant (each contour is a *conductor*).

Example 4 A condenser consists of two plates in the form of coplanar half planes with parallel edges separated by the distance $2a$ and with potential difference $2u_0$. Any cross section normal to the planes yields a plane field with two cuts (see Figure 5.17). Again the function

$$w = \frac{2u_0}{\pi} \sin^{-1}\left(\frac{z}{a}\right)$$

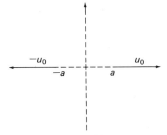

Figure 5.17

maps the domain onto the strip $|u| < u_0$. Thus the equipotential lines are the hyperbolas

$$\frac{x^2}{a^2 \sin^2 \frac{\pi u}{2u_0}} - \frac{y^2}{a^2 \cos^2 \frac{\pi u}{2u_0}} = 1.$$

We gather all these analogies in Table 5.1. Similar analogies with fluid flow are available for steady state diffusion, static magnetic and gravitational fields, and hydromechanics.

Table 5.1

Analogies of Fluid Flow, Heat Flow, and Electrostatic Fields

	Fluid flow	Heat flow	Electrostatic field
Complex potential	$w(z) = u + iv$	$w(z) = u + iv$	$iw(z) = -v + iu$
Vector field	$V = \overline{w'(z)}$ = grad u	$Q = -k\overline{w'(z)}$ = $-k$ grad u	$E = -\overline{w'(z)}$ = $-$ grad u
u $u(z) = $ constant v	Potential function Equipotential lines Stream function	Temperature Isotherms Stream function	Potential function Equipotential lines $-v$ is the force function
$v(z) = $ constant	Stream lines	Stream lines	Lines of force

EXERCISES

1. Find the equation of the stream lines for a flow of an incompressible fluid without sources or sinks in the following domains:

(a) $|\arg z| < \dfrac{\pi}{4}$,

(b) $|\arg z| < \dfrac{\pi}{8}$,

(c) $|\arg z| < \dfrac{3\pi}{4}$,

(d) $|x| < 1, \quad y > 0$.

2. Find the stream lines for a dam of height a if the flow is infinitely deep and has velocity $A > 0$ as $z \to \infty$. What is the velocity at 0 and ia (see Figure 5.18)?

5.5 PHYSICAL APPLICATIONS 143

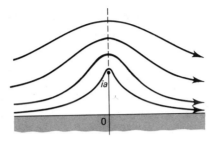

Figure 5.18

3. Find the complex potential and stagnation points for a flow of an incompressible fluid through the domain in Figure 5.19 assuming the original velocity of the flow is A.

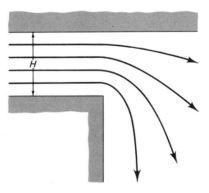

Figure 5.19

4. Find the isotherms of the plate indicated in Figure 5.20 with temperature $0°$ on the horizontal side and $1°$ on the vertical sides.

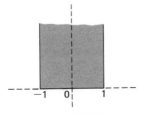

Figure 5.20

5. Find the isotherms in the infinite slab $0 < y < \pi$, if the edges are insulated for $x < 0$ and the temperature satisfies $u(x) = 0°$ and $u(x + i\pi) = 1°$, for $x \geq 0$.
6. A condenser consists of three parallel plates: the center one a half plane, the others, planes with cross section and potential as indicated in Figure 5.21. Find an expression for the equipotential lines. (*Hint*: Use the Schwarz–Christoffel mapping.)

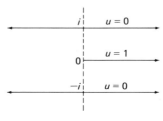

Figure 5.21

NOTES

Section 5.1 A table of elementary conformal mappings is given in the appendix. Further mappings may be found in [Ko].

Section 5.2 Linear fractional transformations are also known as linear transformations, bilinear transformations, linear substitutions, and Möbius transformations.

Section 5.4 Two different proofs of the Riemann Mapping Theorem may be found in [V]. See [A, pp. 227–232] for a proof of the Schwarz–Christoffel formula. A formula for mapping the unit disk onto the *exterior* of a polygon is easily derived.

Section 5.5 Problems involving sources will be discussed in the next chapter. Some detailed application may be found in [R].

Chapter 6 | BOUNDARY-VALUE PROBLEMS

6.1 HARMONIC FUNCTIONS

Let $f(z) = u(z) + iv(z)$ be analytic in the domain G. Since $f'(z) = u_x + iv_x = v_y - iu_y$ is analytic, the second partial derivatives of u and v are continuous and satisfy

$$u_{xx} = (v_y)_x = -u_{yy}, \qquad v_{xx} = (-u_y)_x = -v_{yy},$$

by the Cauchy–Riemann equations. Thus, the real functions u and v both satisfy *Laplace's equation*

$$\Delta u = u_{xx} + u_{yy} = 0, \qquad \Delta v = v_{xx} + v_{yy} = 0,$$

and are said to be *harmonic functions*. A function v is called a *harmonic conjugate* of u whenever $f = u + iv$ is analytic. Note that $-u$ is a harmonic conjugate of v since $-if = v - iu$. Given a harmonic function in a simply connected domain, we may construct its conjugates as follows:

(i) Integrate u_x with respect to y obtaining

$$v = \int u_x \, dy + C(x).$$

(ii) Evaluate $C'(x)$ by equating v_x with $-u_y$, and integrate to obtain $C(x)$.

Example 1 The function $u = \sin x \cosh y$ is harmonic since

$$\Delta u = u_{xx} + u_{yy} = -u + u = 0.$$

Then

(i) $v = \int \cos x \cosh y \, dy + C(x) = \cos x \sinh y + C(x)$,
(ii) $-\sin x \sinh y = -u_y = v_x = -\sin x \sinh y + C'(x)$,

so $C'(x) = 0$ and C is constant. Hence $v = \cos x \sinh y + C$, and

$$f(z) = u + iv = \sin z + C.$$

The correspondence between analytic and harmonic functions yields many important properties for the latter:

Maximum Principle If $u(z)$ is harmonic and nonconstant in a simply connected domain G, then $u(z)$ has no maximum or minimum in G.

Proof Constructing a conjugate harmonic function $v(z)$, we have that $f = u + iv$ is analytic in G. Likewise

$$F(z) = e^{f(z)} = e^{u+iv}$$

is analytic in G, and $|F(z)| = e^{u(z)}$. Since $F(z)$ is nonzero in G, applying the Maximum and Minimum Principles for analytic functions to F, it follows that e^u has no maximum or minimum in G. Since the real function e^u is an increasing function of u, the proof is complete.

Mean-Value Theorem If $u(z)$ is harmonic in $|z - \zeta| < R$, then

$$u(\zeta) = \frac{1}{2\pi} \int_0^{2\pi} u(\zeta + re^{i\theta}) \, d\theta, \qquad 0 < r < R.$$

6.1 HARMONIC FUNCTIONS

Proof This equation is merely the real part of Gauss's Mean-Value Theorem (Exercise 4, Section 2.4) which follows directly from the Cauchy Integral Formula. Moreover,

$$\frac{1}{\pi R^2} \iint_{|z-\zeta|<R} u(z) r \, dr \, d\theta = \frac{2u(\zeta)}{R^2} \int_0^R r \, dr = u(\zeta),$$

yielding the *Area Mean-Value Theorem*. The most important point to notice is that the boundary values determine the value at the center of the disk, leading us to ask whether some analog of the Cauchy Integral Formula exists for harmonic functions. We can generalize further and pose *Dirichlet's Problem*: Given an *arbitrary* domain G, is there a function harmonic in G having preassigned values on the boundary? Solutions would have immediate physical applications in streamlining, as this was the situation studied in Section 5.5. The existence of a solution depends on the *geometry* of the domain G, but a solution continuous on the closure of G is necessarily unique (see Exercises 7 and 9).

If $w = g(z)$ maps the closure of G conformally onto $|w| \leq 1$, $g(z_0) = 0$, and $u(z)$ is harmonic on the closure of G, then $u \circ g^{-1}$ is harmonic on $|w| \leq 1$, so by the Mean-Value Theorem, for $w = e^{i\theta}$ and $dw = iw \, d\theta$,

$$u(z_0) = u \circ g^{-1}(0) = \frac{1}{2\pi i} \int_{|w|=1} u \circ g^{-1}(w) \frac{dw}{w}$$

$$= \frac{1}{2\pi i} \int_{\partial G} u(z) \frac{g'(z)}{g(z)} \, dz.$$

Example 2 If $G = \{z : |z| < R\}$, then the required mapping is

$$g(z) = \frac{R(z - z_0)}{R^2 - \bar{z}_0 z}.$$

But, for $|z| = R$,

$$\frac{g'(z)}{g(z)} = \frac{R^2 - |z_0|^2}{(z - z_0)(R^2 - \bar{z}_0 z)} = \frac{|z|^2 - |z_0|^2}{z|z - z_0|^2} = \frac{1}{z} \operatorname{Re}\left(\frac{z + z_0}{z - z_0}\right),$$

and the expression becomes *Poisson's Integral Formula* for the disk

$$u(z_0) = \frac{1}{2\pi} \int_0^{2\pi} u(Re^{i\phi}) \operatorname{Re}\left(\frac{Re^{i\phi} + z_0}{Re^{i\phi} - z_0}\right) d\phi.$$

Since z_0 is arbitrary, the formula holds for all $|z_0| < R$. We shall consider the formula in greater detail in the next section.

EXERCISES

1. If u and v are harmonic functions, show that $au + bv$ and $\max(u, v)$ are also harmonic, where a and b are real constants. Show that uv is harmonic if u and v are conjugate harmonic functions.
2. Find the conjugates of the following harmonic functions:
   (a) $u = x^2 - (y-1)^2$,
   (b) $u = \frac{1}{2} \log(x^2 + y^2)$,
   (c) $u = \tan^{-1} \frac{2xy}{x^2 - y^2}$,
   (d) $u = \frac{x(x-1) + y^2}{(x-1)^2 + y^2}$.
3. Use the Area Mean-Value Theorem to prove the Maximum Principle for harmonic functions.
4. Show $\log |f(z)|$ is harmonic whenever $f(z)$ is analytic and nonzero.
5. If u is harmonic in a simply connected domain G, show $\overline{\operatorname{grad} u}$ is analytic. Then, using the development in Section 5.5, prove
$$\int_\gamma u_n \, ds = 0,$$
where γ is any pwd closed curve in G and u_n is the directional derivative of u in the outward normal direction.
6. Show the Maximum Principle holds for multiply connected domains.
7. Prove any solution of Dirichlet's Problem continuous on the closure of the domain must be unique. (*Hint*: Use the Maximum Principle of Exercise 6).
8. Suppose $u(z)$ and $v(z)$ are harmonic on a domain G, continuous on its closure, and satisfy $u(z) \leq v(z)$ on the boundary of G. Show $u(z) \leq v(z)$ for every z in G.
9. Let G be the domain $0 < |z| < 1$. Show there is no function $u(z)$ harmonic in G with boundary values $u(e^{i\theta}) = 0$, $u(0) = a > 0$. (*Hint*: Apply Exercise 8 to the functions
$$u_r(z) = a \frac{\log |z|}{\log r},$$
harmonic in $r < |z| < 1$.)
10. Prove $\int_0^\pi \log \sin \theta \, d\theta = -\pi \log 2$. (*Hint*: Apply the Mean-Value Theorem to $\log |1 + z|$ in $|z| \leq r < 1$, and let $r \to 1^-$.)

6.2† POISSON'S INTEGRAL FORMULA

In the applications of Section 5.5, the given boundary values were not necessarily continuous. It is of interest to discover whether the integral formula yields a harmonic function in spite of the discontinuities. This leads to the following theorem:

6.2 POISSON'S INTEGRAL FORMULA

Poisson's Integral Formula Let $U(\phi)$ be continuous for $0 \le \phi \le 2\pi$ except for a finite number of jumps. Then the function

$$u(z) = \frac{1}{2\pi} \int_0^{2\pi} U(\phi) \, \text{Re}\left(\frac{Re^{i\phi} + z}{Re^{i\phi} - z}\right) d\phi$$

is harmonic in $|z| < R$, and

$$\lim_{z \to Re^{i\phi}} u(z) = U(\phi)$$

at all points of continuity of $U(\phi)$.

Proof We can repeatedly perform partial differentiations with respect to x and y under the integral sign as the resulting integrand is continuous on $|z| \le r < R$. Then

$$\Delta u(z) = \frac{1}{2\pi} \int_0^{2\pi} U(\phi) \, \Delta \text{Re}\left(\frac{Re^{i\phi} + z}{Re^{i\phi} - z}\right) d\phi = 0, \qquad |z| < R,$$

since the real part of the analytic function $(Re^{i\phi} + z)/(Re^{i\phi} - z)$ is harmonic. Thus $u(z)$ is harmonic in $|z| < R$. To prove the remainder of the theorem, note that

$$\text{Re}\left(\frac{Re^{i\phi} + z}{Re^{i\phi} - z}\right) \ge 0, \tag{1}$$

since the analytic function maps the unit disk onto the right half plane,

$$\frac{1}{2\pi} \int_0^{2\pi} \text{Re}\left(\frac{Re^{i\phi} + z}{Re^{i\phi} - z}\right) d\phi = 1, \tag{2}$$

because letting $w = Re^{i\phi}$, $dw = iRe^{i\phi} \, d\phi$, the left side equals

$$\text{Re}\left\{\frac{1}{2\pi i} \int_{|w|=R} \left(\frac{w+z}{w-z}\right) \frac{dw}{w}\right\} = \text{Re}\left\{\text{Res}_0 \frac{w+z}{w(w-z)} + \text{Res}_z \frac{w+z}{w(w-z)}\right\} = 1,$$

$$\frac{R - |z|}{R + |z|} \le \text{Re}\left(\frac{Re^{i\phi} + z}{Re^{i\phi} - z}\right) \le \frac{R + |z|}{R - |z|}, \tag{3}$$

since for $z = re^{i\theta}$ we have

$$\frac{R-r}{R+r} = \frac{R^2 - r^2}{R^2 + r^2 + 2Rr} \le \frac{R^2 - r^2}{R^2 + r^2 - 2Rr\cos(\phi - \theta)} \le \frac{R^2 - r^2}{R^2 + r^2 - 2Rr}$$

$$= \frac{R+r}{R-r},$$

6 BOUNDARY-VALUE PROBLEMS

and the middle term equals

$$\frac{R^2 - r^2}{|Re^{i\phi} - re^{i\theta}|^2} = \frac{\text{Re}[(Re^{i\phi} + re^{i\theta}) \cdot (Re^{-i\phi} - re^{-i\theta})]}{|Re^{i\phi} - re^{i\theta}|^2} = \text{Re}\left(\frac{Re^{i\phi} + z}{Re^{i\phi} - z}\right).$$

If $U(\phi)$ is continuous at $\phi = \alpha$, then given $\varepsilon > 0$, there is a $\delta > 0$ such that $|U(\phi) - U(\alpha)| < \varepsilon$, whenever $|\phi - \alpha| < \delta$ (assume U has period 2π). Then by (2)

$$u(z) - U(\alpha) = \frac{1}{2\pi} \int_0^{2\pi} [U(\phi) - U(\alpha)] \, \text{Re}\left(\frac{Re^{i\phi} + z}{Re^{i\phi} - z}\right) d\phi,$$

so, using (1) and the proof of (3) we have (see Figure 6.1)

$$|u(z) - U(\alpha)| \le \frac{1}{2\pi} \int_0^{2\pi} |U(\phi) - U(\alpha)| \, \text{Re}\left(\frac{Re^{i\phi} + z}{Re^{i\phi} - z}\right) d\phi$$

$$\le \frac{\varepsilon}{2\pi} \int_{|\phi - \alpha| < \delta} \text{Re}\left(\frac{Re^{i\phi} + z}{Re^{i\phi} - z}\right) d\phi$$

$$+ \frac{1}{2\pi} \int_{\delta \le |\phi - \alpha| \le \pi} \frac{R^2 - |z|^2}{|Re^{i\phi} - z|^2} |U(\phi) - U(\alpha)| \, d\phi.$$

Now, if $|\arg z - \alpha| < \delta/2$ and $|\phi - \alpha| \ge \delta$, then

$$|Re^{i\phi} - z| \ge R \sin \frac{\delta}{2},$$

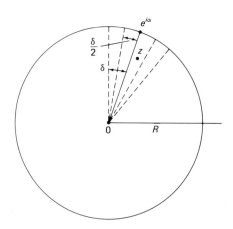

Figure 6.1

6.2 POISSON'S INTEGRAL FORMULA

so by (2) and (3)

$$|u(z) - U(\alpha)| \leq \varepsilon + \frac{R^2 - |z|^2}{2\pi R^2 \sin^2 \frac{\delta}{2}} \int_0^{2\pi} |U(\phi) - U(\alpha)| \, d\phi \to \varepsilon,$$

as $z \to Re^{i\alpha}$. Since ε is arbitrary the proof is complete.

As a corollary we have:

Harnack's Inequality If $u(z)$ is harmonic and nonnegative in $|z| < R$, then

$$u(0) \frac{R - |z|}{R + |z|} \leq u(z) \leq u(0) \frac{R + |z|}{R - |z|}.$$

Proof By the Mean-Value Theorem, Poisson's Integral Formula, and (3)

$$u(0) \frac{\rho - |z|}{\rho + |z|} = \frac{1}{2\pi} \int_0^{2\pi} u(\rho e^{i\phi}) \frac{\rho - |z|}{\rho + |z|} \, d\phi \leq u(z), \qquad |z| < \rho < R.$$

Letting $\rho \to R$ gives the first inequality; the second is proved similarly.

EXERCISES

1. Prove the second part of Harnack's Inequality.
2. If $f(z) = u(z) + iv(z)$ is analytic in a domain containing $|z| \leq R$, prove Schwarz's Formula

$$f(z) = \frac{1}{2\pi} \int_0^{2\pi} \frac{Re^{i\phi} + z}{Re^{i\phi} - z} u(Re^{i\phi}) \, d\phi + iv(0).$$

3. Show Schwarz's formula can be rewritten in the form

$$f(z) = \frac{1}{\pi i} \int_{|\zeta| = R} \frac{u(\zeta)}{\zeta - z} \, d\zeta - \overline{f(0)}.$$

[*Hint.* Apply the Mean-Value Theorem to $u(0) = iv(0) + \overline{f(0)}$.]

4. Let $U(\phi)$ be continuous for $0 \le \phi \le 2\pi$ except for a finite number of jumps. Prove

$$g(z) = \frac{z}{\pi} \int_0^{2\pi} \frac{U(\phi)}{Re^{i\phi} - z} d\phi$$

is analytic in $|z| < R$. (*Hint*: Rewrite Schwarz's Formula.)

5. Show, using the method in Section 6.1, that

$$u(z) = \frac{y}{\pi} \int_{-\infty}^{\infty} \frac{u(t)}{(t-x)^2 + y^2} dt, \qquad z = x + iy,$$

is the Poisson Integral Formula for the upper half plane $y > 0$.

6.3 APPLICATIONS

In Section 5.3 we studied three analogous examples of steady state vector fields occuring in nature: fluid flow, heat flow, and the electrostatic field. The vector fields, assumed to be two-dimensional and irrotational, were examined within a domain G containing no sources and sinks. In this section, these problems will be extended to include sources and vortices in the domain. The theory will be developed for fluid flow, and the analogies with the other two fields listed in Table 6.1 at the end of the section.

Recall the velocity vector $V(z)$ of the field equals $\overline{w'(z)}$, where the analytic function $w(z)$ is the complex potential of the flow. Thus the potential function u and the stream function v are conjugate harmonic functions, and the problem of finding the streamlines reduces to Dirichlet's Problem. Observe by the Maximum Principle, that *if an equipotential line forms a closed curve γ, then either γ encloses a singularity of $u(z)$ or $u(z)$ is constant in G.* Of course, *there are no closed streamlines*, since the flow is irrotational. Furthermore, *neither streamlines nor equipotential lines begin or end at an interior point z_0 of G*, otherwise a sufficiently small disk centered at z_0 lies in G and its boundary meets the streamline $v(z) = k$ at only one point. By continuity, the remaining boundary points satisfy either $v(z) > k$ or $v(z) < k$, violating the Mean-Value Theorem. Thus distinct streamlines can meet only at boundary points of G (for example, at sources) or at infinity. We illustrate with a problem in heat flow:

Example 1 Let the plate G be disk of radius R with boundary temperatures $1°$ in the upper half plane and $0°$ in the lower half plane. Find the temperature at all points of G and describe the isotherms.

6.3 APPLICATIONS

Applying the Poisson Integral Formula we have for $z = re^{i\theta}$, $r < R$,

$$u(z) = \frac{1}{2\pi} \int_0^\pi \text{Re}\left(\frac{Re^{i\phi} + z}{Re^{i\phi} - z}\right) d\phi$$

$$= \frac{1}{2\pi} \int_0^\pi \frac{R^2 - r^2}{R^2 + r^2 - 2Rr\cos(\phi - \theta)} d(\phi - \theta)$$

$$= \frac{1}{\pi} \tan^{-1}\left(\frac{R+r}{R-r} \tan \frac{\phi - \theta}{2}\right)\bigg|_0^\pi.$$

Thus, since $\cot \theta/2 = \tan(\pi - \theta)/2$,

$$\tan \pi u(z) = \frac{\frac{R+r}{R-r}\left(\tan \frac{\pi - \theta}{2} + \tan \frac{\theta}{2}\right)}{1 - \left(\frac{R+r}{R-r}\right)^2 \tan \frac{\pi - \theta}{2} \tan \frac{\theta}{2}}$$

$$= \frac{R^2 - r^2}{-2Rr \sin \theta}$$

$$= \tan\left(\text{Arg } i\frac{R+z}{R-z}\right),$$

implying that the temperature is given by

$$u(z) = \frac{1}{\pi} \text{Arg } i \frac{R+z}{R-z}.$$

The isotherms satisfy

$$\text{Arg } i \frac{R+z}{R-z} = \text{constant},$$

and $i(R + z)/(R - z)$ maps $|z| < R$ onto the upper half plane, so the isotherms correspond to the arcs of the family of circles passing through the points $\pm R$ lying in $|z| < R$ (see Figure 6.2).

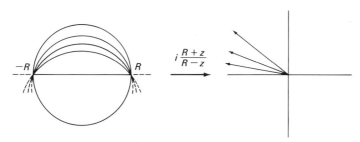

Figure 6.2

Assume a *point source* (or sink) is located at the origin. Then the flow Q across any Jordan curve around the origin is a nonzero constant, and if γ is the circle $|z| = r$, the normal velocity component V_n is constant in each direction, as the streamlines are radial at the origin. Thus

$$Q = \int V_n \, ds = V_n \int_0^{2\pi} r \, d\theta = 2\pi r V_n,$$

and

$$V(z) = V_n \cdot \frac{z}{|z|} = \frac{Q}{2\pi} \frac{z}{|z|^2},$$

since $z/|z|$ is the unit normal vector. Since $V(z) = \overline{w'(z)}$,

$$w(z) = \frac{Q}{2\pi} \log(z) + c, \qquad c \text{ complex},$$

hence the potential function and stream function are given by

$$u(z) = \frac{Q}{2\pi} \log |z|, \qquad v(z) = \frac{Q}{2\pi} \arg z,$$

respectively, up to an arbitrary real constant. Note that $v(z)$ is multivalued, and both functions are harmonic in any Jordan domain not containing the origin. If $Q > 0$, we have a source of *strength* Q at $z = 0$, and if $Q < 0$, we have a sink.

If the source is not at the origin, but at the point z_0, the complex potential is

$$w(z) = \frac{Q}{2\pi} \log(z - z_0) + c. \tag{1}$$

On the other hand, the vector field may not be irrotational. This may occur, for example, from the action of a cylindrical rotor, so that in any plane normal to its axis, the streamlines are concentric circles centered on the rotor. Such a field is called a *plane vortex field*.

If a *point vortex* is located at the origin, the circulation Γ along any Jordan curve γ is a nonzero constant ($\Gamma > 0$ when the flow is counterclockwise). Along a circle $|z| = r$, the tangential velocity component V_s is constant, so

$$\Gamma = \int_\gamma V_s \, ds = 2\pi r V_s,$$

and

$$V(z) = V_s \cdot \frac{iz}{|z|} = \frac{i\Gamma}{2\pi} \frac{z}{|z|^2},$$

6.3 APPLICATIONS

since $iz/|z|$ is the unit tangential vector. Then, except for an arbitrary constant,

$$w(z) = \frac{-i\Gamma}{2\pi} \log z = \frac{\Gamma}{2\pi} \arg z + i \frac{\Gamma}{2\pi} \log \frac{1}{|z|} \qquad (2)$$

is the complex potential of this field. As a point source may also be a vortex, we combine equations (1) and (2) (at z_0) obtaining

$$w(z) = \frac{\Gamma + iQ}{2\pi i} \log(z - z_0) + c, \qquad (3)$$

as the complex potential of a *vortex-source* located at z_0 with *intensity* Γ and *strength* Q. The complex potential of a system of vortex sources $\Gamma_1 + iQ_1$, ..., $\Gamma_k + iQ_k$ located at $z_1, \ldots, z_k$ is obtained by adding up the individual complex potentials

$$w(z) = \frac{1}{2\pi i} \sum_{j=1}^{k} (\Gamma_j + iQ_j) \log(z - z_j), \qquad (4)$$

as the vector field is obtained by superposition. Furthermore, this result and the usual limiting procedure may be employed to obtain the complex potential of a line L of sources, provided the flow function $Q(\zeta)$ is integrable

$$w(z) = \frac{1}{2\pi} \int_L Q(\zeta) \log(z - \zeta) \, ds, \qquad \zeta \text{ on } L. \qquad (5)$$

Two examples are of particular interest:

Example 2 If the system consists of *two sources*, each of strength Q, located at z_1 and z_2, the complex potential is given by

$$w(z) = \frac{Q}{2\pi} \log(z - z_1)(z - z_2).$$

The equipotential lines, satisfying

$$|z - z_1| \, |z - z_2| = \text{constant},$$

are known as *lemniscates* and are shown in Figure 6.3. The lemniscate shaped like ∞, is given by the equation

$$|z - z_1| \, |z - z_2| = \frac{|z_1 - z_2|^2}{4},$$

and $(z_1 + z_2)/2$ is a stagnation point.

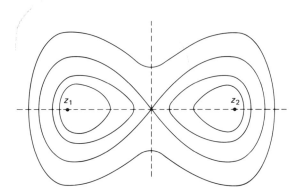

Figure 6.3
Lemniscates.

Example 3 A system consisting of a *source* and a *sink* of strengths Q and $-Q$ situated at z_1 and z_2, respectively, has a complex potential given by the equation

$$w(z) = \frac{Q}{2\pi} \log \frac{z - z_1}{z - z_2}.$$

The equipotential lines satisfy

$$\left| \frac{z - z_1}{z - z_2} \right| = \text{constant},$$

and form the Circles of Apollonius indicated as solid lines in Figure 6.4 and the streamlines are the family of circles passing through z_1 and z_2.

Let $z_1 = -h$, $z_2 = 0$, then

$$w(z) = \frac{Q}{2\pi} \log \frac{z + h}{z} = \frac{p}{2\pi} \log\left(1 + \frac{h}{z}\right)^{1/h}, \qquad p = Qh.$$

If we now permit the source to approach the sink simultaneously increasing Q so that p remains constant, we obtain in the limit a *point doublet of moment p* located at 0 whose streamlines are directed along the positive real axis. Its complex potential is given by

$$w(z) = \frac{p}{2\pi} \lim_{h \to 0} \log\left(1 + \frac{h}{z}\right)^{1/h} = \frac{p}{2\pi} \log e^{1/z} = \frac{p}{2\pi z}, \tag{6}$$

6.3 APPLICATIONS

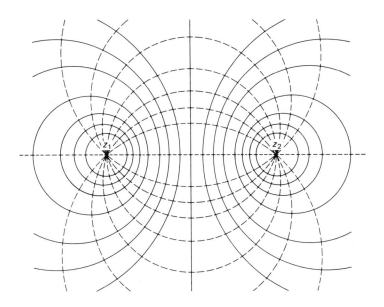

Figure 6.4

Circles of Apollonius.

hence

$$u = \frac{p}{2\pi} \frac{x}{x^2 + y^2}, \qquad v = \frac{-p}{2\pi} \frac{y}{x^2 + y^2}.$$

Then

$$\left(x - \frac{p}{4\pi u}\right)^2 + y^2 = \left(\frac{p}{4\pi u}\right)^2, \qquad x^2 + \left(y + \frac{p}{4\pi v}\right)^2 = \left(\frac{p}{4\pi v}\right)^2,$$

and the equipotential lines and streamlines are the families of circles indicated in Figure 6.5.

The procedure above also holds for z_1 complex, but now the moment of the doublet is complex with argument $\pi + \arg z_1$ coinciding with the direction of the streamlines at the origin.

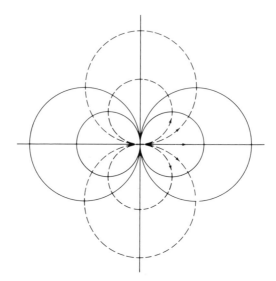

Figure 6.5
Point doublet (dipole) at the origin.

Example We consider the problem of complete streamlining of the exterior of the unit disk so that the velocity vector tends to 1 at ∞.

As was shown in Example 1, Section 5.5, if the flow is symmetric with the x-axis, the complex potential is given by

$$w_1(z) = z + \frac{1}{z},$$

since $V_1(z) = 1 - (1/\bar{z}^2)$. Dropping the assumption of symmetry, observe that the flow might also be subject to a vortex flow centered at the origin of intensity Γ, with complex potential

$$w_2(z) = \frac{\Gamma}{2\pi i} \log z,$$

since its corresponding velocity vector

$$V_2(z) = \frac{i\Gamma}{2\pi \bar{z}}$$

vanishes at ∞. By superposition, the equation of the complex potential is given by

$$w(z) = z + \frac{1}{z} + \frac{\Gamma}{2\pi i} \log z.$$

6.3 APPLICATIONS

The magnitude of the velocity satisfies

$$|\overline{V(z)}| = |w'(z)| = \left|1 - \frac{1}{z^2} + \frac{\Gamma}{2\pi i z}\right|,$$

vanishing at the zeros z_s (stagnation points) of the equation

$$z^2 + \frac{\Gamma}{2\pi i} z - 1 = 0,$$

That is,

$$z_s = \frac{\Gamma i \pm \sqrt{16\pi^2 - \Gamma^2}}{4\pi}. \qquad (7)$$

If $|\Gamma| < 4\pi$, then $|z_s| = \sqrt{\Gamma^2 + 16\pi^2 - \Gamma^2}/4\pi = 1$ and

$$\tan \mathrm{Arg}\, z_s = \frac{\pm \Gamma}{\sqrt{16\pi^2 - \Gamma^2}},$$

and if $|\Gamma| > 4\pi$, the stagnation points are on the imaginary axis and satisfy

$$|z_s| = \frac{\Gamma \pm \sqrt{\Gamma^2 - 16\pi^2}}{4\pi}.$$

Thus only one of them is outside the unit circle. A sketch of the streamlines is shown in Figure 6.6.

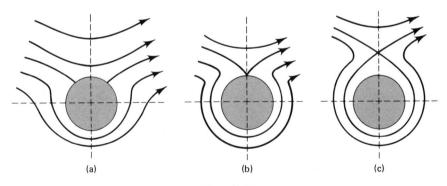

Figure 6.6

Complete streamlining of the exterior of a disk with a point vortex at its center.
(a) $0 < \Gamma < 4\pi$. (b) $\Gamma = 4\pi$. (c) $\Gamma > 4\pi$.

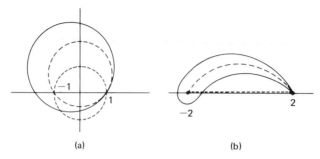

Figure 6.7

Joukowsky profile. (a) ζ-plane. (b) z-plane.

To completely streamline a domain G, we need merely map G conformally onto the exterior of the unit disk, $f\colon G \to \{|z| > 1\}$, then the composite function $w \circ f$ is the complex potential of G. Of particular interest in aerodynamics is the complete streamlining of the *Joukowsky profile*, given by the mapping $z = \zeta + 1/\zeta$, mapping given circles as shown in Figure 6.7. The profiles can be made to approximate cross sections of airfoils, and the lift of the airfoil may be evaluated.

Table 6.1

Steady State Vector Fields

Fluid flow	Heat flow	Electrostatic field
$w(z) = \dfrac{\Gamma + iQ}{2\pi i} \log(z - z_0)$	$w(z) = \dfrac{Q}{2\pi k} \log \dfrac{1}{z - z_0}$	$iw(z) = \dfrac{Q}{2\pi} i \log \dfrac{1}{z - z_0}$
Vortex source of strength Q and intensity Γ at z_0	*Source* of strength Q at z_0	*Charge* of magnitude $Q/2\pi$ at z_0
$w(z) = \dfrac{p}{2\pi} \dfrac{1}{z - z_0}$	$w(z) = \dfrac{-p}{2\pi k} \dfrac{1}{z - z_0}$	$iw(z) = \dfrac{-ip}{2\pi} \dfrac{1}{z - z_0}$
Doublet of moment p at z_0	*Doublet* of moment p at z_0	*Dipole* of moment $p/2\pi$ at z_0

EXERCISES

1. Find the potential function for the plane electrostatic field in $|z| < 1$ bordered by electrodes represented by the semicircles $e^{i\theta}$, $|\theta| < \pi/2$ and $e^{i\theta}$, $|\theta - \pi| < \pi/2$, with potentials of u_0 and u_1 respectively.

6.3 APPLICATIONS

2. Find the temperature of a plate Q in the shape of the upper half plane given boundary temperatures of $100°$ on $|x| > 1$ and $0°$ on $|x| < 1$.
3. Find the complex potential and the streamlines for the plane flow of a fluid in the upper half plane when there is a source of strength Q at i and a sink of equal strength at 0.
4. What is the complex potential for the plane flow of a fluid with a sink of strength Q at -1 and a vortex source of strength Q and intensity Γ at 0?

In Exercise 5–8, given the complex potential of the flow of a fluid, construct the equipotential lines and streamlines, and find the velocity vector V, the stagnation points, the strength and intensity of the vortex sources, the moments of the doublets, and the behavior of the flow at ∞.

5. $w(z) = \dfrac{Q}{2\pi} \log\left(z + \dfrac{1}{z}\right).$

6. $w(z) = \log\left(1 + \dfrac{4}{z^2}\right).$

7. $w(z) = \log\left(z^2 - \dfrac{1}{z^2}\right).$

8. $w(z) = az + \dfrac{Q}{2\pi} \log \dfrac{1}{z}, \quad a, Q > 0.$

9. A point *multiplet* (multipole) is a generalization of a doublet (dipole), obtained by taking a sink of strength Q at the origin together with n sources of strength Q/n symmetrically distributed on a circle of radius r, and holding Qr fixed as r tends to 0. Show that its complex potential is given by

$$w(z) = \dfrac{-p}{2\pi n} \dfrac{1}{z^n}, \quad |p| = Qr,$$

and that the streamlines are directed along the arguments of the nth roots of p. Such a multiplet is said to have *order* $2n$.

10. Sketch the image of the circles

    (a) $|\zeta - i| = \sqrt{2},$ (b) $|\zeta + 1 - i| = \sqrt{5},$

    under the mapping $z = \zeta + (1/\zeta)$. (*Hint*: Show that

    $$\dfrac{z-2}{z+2} = \left(\dfrac{\zeta-1}{\zeta+1}\right)^2.)$$

NOTES

Section 6.1   We have assumed the domain G is simply connected in order to force the conjugate harmonic function to be single-valued. For multiply connected domains, the conjugate harmonic function may be many-valued [see Exercise 2(b)]. The construction is essentially the same and the Maximum Principle follows with minor modifications. A more thorough discussion of Dirichlet's Problem may be found in [A, pp. 237–253].

Section 6.2   The hypothesis on $U(\phi)$ in Poisson's Integral Formula may be relaxed substantially [Hf, Chapter 3 and H, Chapter 19].

Section 6.3   A complete treatment of the Joukowski profile may be found in [R, pp. 115–121].

Chapter 7 | FOURIER AND LAPLACE TRANSFORMATIONS

7.1 FOURIER SERIES

The Poisson Integral Formula is intimately related to the notion of Fourier series. We have seen that if $U(\phi)$ is piecewise continuous for $0 \leq \phi \leq 2\pi$, then

$$u(z) = \frac{1}{2\pi} \int_0^{2\pi} U(\phi) \, \text{Re}\left(\frac{Re^{i\phi} + z}{Re^{i\phi} - z}\right) d\phi$$

is harmonic in $|z| < R$ and has boundary values $u(Re^{i\phi}) = U(\phi)$ at all points of continuity of U. In practice, it is often easiest to obtain $u(z)$ by expanding the right side of the above equation into an infinite series. Then

$$u(z) = \text{Re}\left[\frac{1}{2\pi}\int_0^{2\pi} U(\phi)\left[1 + 2\sum_{n=1}^{\infty}\left(\frac{z}{Re^{i\phi}}\right)^n\right]d\phi\right],$$

and since the series converges uniformly in $|z| \leq \rho < R$, term-by-term integration is permissible, yielding the *Fourier series*

$$u(z) = \frac{1}{2\pi}\int_0^{2\pi} U(\phi)\,d\phi + 2\,\text{Re}\left[\sum_{n=1}^{\infty}\left(\frac{1}{2\pi}\int_0^{2\pi} U(\phi)e^{-in\phi}\,d\phi\right)\left(\frac{z}{R}\right)^n\right]$$

$$= c_0 + 2\,\text{Re}\left[\sum_{n=1}^{\infty} r^n c_n e^{in\theta}\right], \qquad z = re^{i\theta}, \tag{1}$$

where

$$R^n c_n = \frac{1}{2\pi}\int_0^{2\pi} U(\phi)e^{-in\phi}\,d\phi$$

is the *n*th *complex Fourier coefficient* of $U(\phi)$.

Example 1 If $U(\phi) = \cos\phi$, then

$$2\pi R^n c_n = \int_0^{2\pi} \cos\phi\, e^{-in\phi}\,d\phi$$

$$= \int_0^{2\pi} \frac{e^{-i(n-1)\phi} + e^{-i(n+1)\phi}}{2}\,d\phi$$

$$= \begin{cases} \dfrac{-1}{2i}\left[\dfrac{e^{-i(n-1)\phi}}{n-1} + \dfrac{e^{-i(n+1)\phi}}{n+1}\right]_0^{2\pi} = 0, & n \neq 1, \\ \dfrac{1}{2}\left[\phi - \dfrac{e^{-2i\phi}}{2i}\right]_0^{2\pi} = \pi, & n = 1. \end{cases}$$

Thus $u(z) = \text{Re}(z/R)$.

A similar connection exists between the Fourier series and the Laurent series of a function $f(z)$ analytic in an annulus $r_1 < |z| < r_2$. Here

$$f(z) = \sum_{n=-\infty}^{\infty} c_n z^n, \tag{2}$$

7.1 FOURIER SERIES

where

$$R^n c_n = \frac{R^n}{2\pi i} \int_{|z|=R} \frac{f(z)}{z^{n+1}} dz$$

$$= \frac{1}{2\pi} \int_0^{2\pi} f(Re^{i\phi}) e^{-in\phi} d\phi, \qquad r_1 < R < r_2.$$

Observe that

$$\int_0^{2\pi} \left| \sum_{n=-k}^{k} r^n c_n e^{in\phi} \right|^2 d\phi = \int_0^{2\pi} \left(\sum_{n=-k}^{k} \sum_{m=-k}^{k} r^{n+m} c_n \bar{c}_m e^{i(n-m)\phi} \right) d\phi$$

$$= 2\pi \sum_{n=-k}^{k} r^{2n} |c_n|^2,$$

since

$$\int_0^{2\pi} e^{i(n-m)\phi} d\phi = \frac{e^{i(n-m)\phi}}{i(n-m)} \bigg|_0^{2\pi} = 0, \qquad m \neq n.$$

Since the Laurent series representation converges uniformly in $r_1 < \rho_1 \leq |z| \leq \rho_2 < r_2$, we can interchange limits and integrals obtaining *Parseval's Identity*

$$\int_0^{2\pi} |f(re^{i\phi})|^2 d\phi = \lim_{k \to \infty} \int_0^{2\pi} \left| \sum_{n=-k}^{k} r^n c_n e^{in\phi} \right|^2 d\phi$$

$$= 2\pi \sum_{n=-\infty}^{\infty} r^{2n} |c_n|^2.$$

Note that if $z = e^{i\phi}$, the series in equations (1) and (2) may both be written in the form

$$\sum_{n=-\infty}^{\infty} c_n e^{in\phi}, \tag{3}$$

by letting $c_{-n} = \bar{c}_n$, since $2 \operatorname{Re}(c_n e^{in\phi}) = c_n e^{in\phi} + \bar{c}_n e^{-in\phi}$. In equation (2) the Fourier series converges (uniformly) to $f(e^{i\phi})$ by Laurent's Theorem, but it need not converge to $U(\phi)$ in equation (1). Consider, for example, a function $U_1(\phi)$ differing from $U(\phi)$ at only one point. Both functions have the same Fourier series, hence it cannot represent them both at every point. In fact, continuous functions exist whose Fourier series diverge at all rational numbers ϕ in the interval $[0, 2\pi)$. The problem of convergence is of fundamental importance in the study of Fourier series.

Before studying the question of convergence it is useful to define one-sided limits and derivatives. For $\varepsilon > 0$, the limits

$$U(\phi + 0) = \lim_{\varepsilon \to 0} U(\phi + \varepsilon), \qquad U(\phi - 0) = \lim_{\varepsilon \to 0} U(\phi - \varepsilon)$$

are the right and left side limits, and

$$U'(\phi + 0) = \lim_{\varepsilon \to 0} \frac{U(\phi + \varepsilon) - U(\phi + 0)}{\varepsilon},$$

$$U'(\phi - 0) = \lim_{\varepsilon \to 0} \frac{U(\phi - \varepsilon) - U(\phi - 0)}{\varepsilon},$$

are the right and left side derivatives of the function U at ϕ, respectively. Observe that if U is continuous at ϕ, both one-sided limits coincide with $U(\phi)$ and, if U is differentiable at ϕ, both one-sided derivatives agree with $U'(\phi)$.

A function $U(\phi)$ is said to be *piecewise smooth* on $[a,b]$ if it has a continuous derivative at all but finitely many points at which the one-sided limits and derivatives of U exist.

Our next theorem settles the convergence problem for a useful class of functions:

Theorem Let $U(\phi)$ be a piecewise smooth function on $[0, 2\pi]$ with period 2π, and let

$$c_n = \frac{1}{2\pi} \int_0^{2\pi} U(\theta) e^{-in\theta} \, d\theta.$$

Then

$$\lim_{k \to \infty} \sum_{n=-k}^{k} c_n e^{in\phi} = \tfrac{1}{2}[U(\phi + 0) + U(\phi - 0)].$$

[Note that if the Fourier series (3) converges, it agrees with the limit above, but the latter exists even when (3) diverges.]

Proof If $U(\phi)$ is differentiable on $a < \phi < b$,

$$\int_a^b U(\phi) e^{ik\phi} \, d\phi = \frac{U(\phi) e^{ik\phi}}{ik} \bigg|_a^b - \frac{1}{ik} \int_a^b U'(\phi) e^{ik\phi} \, d\phi$$

7.1 FOURIER SERIES

vanishes as $k \to \infty$, since the last integral is bounded. Thus, the integral over the interval $[0, 2\pi]$ will also vanish as $k \to \infty$. Now

$$\begin{aligned} s_k(\phi) &= \sum_{n=-k}^{k} c_n e^{in\phi} \\ &= \frac{1}{2\pi} \int_0^{2\pi} U(\theta) \left[\sum_{n=-k}^{k} e^{in(\phi-\theta)} \right] d\theta \\ &= \frac{1}{2\pi} \int_0^{2\pi} U(\theta) \left[\frac{e^{-ik(\phi-\theta)} - e^{i(k+1)(\phi-\theta)}}{1 - e^{i(\phi-\theta)}} \right] d\theta \\ &= \frac{1}{2\pi} \int_0^{2\pi} \frac{\sin(k+\tfrac{1}{2})(\phi-\theta)}{\sin \tfrac{1}{2}(\phi-\theta)} U(\theta)\, d\theta, \end{aligned}$$

so setting $t = \theta - \phi$, integrating over the interval $[-\pi, \pi]$, and dividing the range of integration into halves, we can write

$$s_k(\phi) = \frac{1}{2\pi} \int_0^\pi \frac{\sin(k+\tfrac{1}{2})t}{\sin \tfrac{1}{2}t} [U(\phi+t) + U(\phi-t)]\, dt$$

(see Figure 7.1).

In particular, if $U(\phi) = 1$ for all ϕ, then $c_0 = 1$ and $c_n = 0$ for $n \neq 0$, so

$$1 = \frac{1}{2\pi} \int_0^\pi \frac{\sin(k+\tfrac{1}{2})t}{\sin \tfrac{1}{2}t} \cdot 2\, dt.$$

Multiplying this identity by $[U(\phi+0) + U(\phi-0)]/2$ we have

$$s_k(\phi) - \frac{U(\phi+0) + U(\phi-0)}{2}$$

$$= \frac{1}{2\pi} \int_0^\pi \frac{\sin(k+\tfrac{1}{2})t}{\sin \tfrac{1}{2}t} [U(\phi+t) - U(\phi+0) + U(\phi-t) - U(\phi-0)]\, dt.$$

(4)

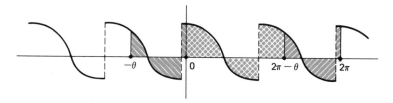

Figure 7.1

Since the one-sided derivatives of U exist, the function

$$\frac{t}{\sin \tfrac{1}{2}t}\left[\frac{U(\phi+t)-U(\phi+0)}{t}+\frac{U(\phi-t)-U(\phi-0)}{t}\right]$$

is piecewise smooth for $0 \leq t \leq \pi$, so the first remark in this proof applies and integral (4) vanishes as $k \to \infty$. Thus

$$\tfrac{1}{2}[U(\phi+0)+U(\phi-0)] = \lim_{k\to\infty} s_k(\phi).$$

Example 2 Let $U(\phi)$ equal $\sin \phi$, for $0 \leq \phi \leq \pi$, and vanish on $\pi \leq \phi \leq 2\pi$. Then $U(\phi)$ is piecewise smooth and

$$2\pi c_n = \int_0^\pi \sin \phi e^{-in\phi}\, d\phi = \frac{1}{2i}\int_0^\pi e^{i(1-n)\phi} - e^{-i(1+n)\phi}\, d\phi,$$

implying $c_{2k} = [\pi(1-4k^2)]^{-1}$, $c_{\pm 1} = \pm[4i]^{-1}$, the remaining coefficients being zero. Since $c_{2k} = c_{-2k}$, $c_{-1} = -c_1$, we have

$$c_{2k}e^{2k\phi i} + c_{-2k}e^{-2k\phi i} = 2c_{2k}\cos 2k\phi,$$
$$c_1 e^{i\phi} + c_{-1}e^{-i\phi} = 2ic_1 \sin \phi,$$

and

$$U(\phi) = \frac{\sin \phi}{2} + \frac{1}{\pi} + \frac{2}{\pi}\sum_{k=1}^\infty \frac{\cos 2k\phi}{(1-2k)(1+2k)}.$$

In particular, for $\phi = \pi/2$, $U(\pi/2) = 1$ so

$$1 = \frac{1}{2} + \frac{1}{\pi} - \frac{2}{\pi}\sum_{k=1}^\infty \frac{(-1)^k}{(2k-1)(2k+1)},$$

hence

$$\frac{\pi-2}{4} = \frac{1}{1\cdot 3} - \frac{1}{3\cdot 5} + \frac{1}{5\cdot 7} - \frac{1}{7\cdot 9} + \cdots.$$

EXERCISES

1. Let the plate G shaped like the unit disk have temperature $u(e^{i\phi}) = \phi$ degrees for $0 \leq \phi < 2\pi$. Show the temperature in G for $z \neq 1$ is given by

$$u(z) = \pi + 2\,\text{Arg}(1-z).$$

7.2 FOURIER TRANSFORMS

2. Find the temperature in $|z| < 1$, given the temperature on the boundary is $u(e^{i\phi}) = \cosh \phi$.
3. Find the Fourier series for $U(\phi) = \pi$ on $0 \leq \phi \leq \pi$ and vanishing on $\pi < \phi < 2\pi$.
4. Find the Fourier series for $U(\phi) = \phi^2$ on $0 \leq \phi < 2\pi$.
5. Use Parseval's Identity to prove Liouville's Theorem. (*Hint*: Show $|c_n| \leq Mr^{-n}$, for all n, where $|f(z)| \leq M$.)
6. Applying Parseval's Identity to the function $U(\phi) = \phi$ show

$$\frac{\pi^2}{6} = \sum_{n=1}^{\infty} \frac{1}{n^2}.$$

7. Show

$$\frac{\pi^4}{90} = \sum_{n=1}^{\infty} \frac{1}{n^4}.$$

8. Applying Parseval's Identity to the function $f(z) = (1 - z)^{-1}$, prove

$$\frac{1}{2\pi} \int_0^{2\pi} \frac{d\phi}{1 - 2r \cos \phi + r^2} = \frac{1}{1 - r^2}, \qquad 0 \leq r < 1.$$

9. Applying Parseval's Identity to the function

$$f(z) = 1 + z + \cdots + z^{n-1},$$

prove

$$\int_0^{2\pi} \left(\frac{\sin \dfrac{n\phi}{2}}{\sin \dfrac{\phi}{2}} \right)^2 d\phi = 2\pi n.$$

(*Hint*: $f(z) = (z^n - 1)/(z - 1)$ for $z \neq 1$.)

7.2 FOURIER TRANSFORMS

The Fourier series of a function $U(\phi)$ of period 2π may be written in the form

$$\sum_{n=-\infty}^{\infty} c_n e^{in\phi}, \qquad c_n = \frac{1}{2\pi} \int_{-\pi}^{\pi} U(\phi) e^{-in\phi} \, d\phi.$$

Similarly, if $U(\phi)$ has period $2\pi\lambda$, setting $\psi = \phi/\lambda$ yields a function of period 2π, hence $U(\phi)$ has the Fourier series

$$\sum_{n=-\infty}^{\infty} c_n e^{in\psi} = \sum_{n=-\infty}^{\infty} c_n e^{in\phi/\lambda},$$

where

$$c_n = \frac{1}{2\pi} \int_{-\pi}^{\pi} U(\lambda\psi) e^{-in\psi}\, d\psi = \frac{1}{2\pi\lambda} \int_{-\pi\lambda}^{\pi\lambda} U(\phi) e^{-in\phi/\lambda}\, d\phi.$$

However, many interesting functions are not periodic, for example, a single unrepeated pulse. We might hope to approximate this situation by a function consisting of identical pulses each a distance $2\pi\lambda$ apart, and investigating the effect on its Fourier series as $\lambda \to \infty$ (see Figure 7.2). Letting $t_n = n/\lambda$, defining

$$u(t) = \frac{1}{\sqrt{2\pi}} \int_{-\pi\lambda}^{\pi\lambda} U(\phi) e^{-it\phi}\, d\phi,$$

and observing $t_{n+1} - t_n = 1/\lambda$, we can write the Fourier series in the form

$$\sum_{n=-\infty}^{\infty} \frac{u(t_n)}{\lambda\sqrt{2\pi}} e^{it_n\phi} = \frac{1}{\sqrt{2\pi}} \sum_{n=-\infty}^{\infty} u(t_n) e^{it_n\phi}(t_{n+1} - t_n),$$

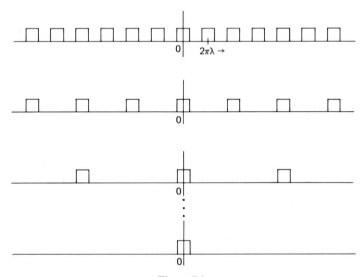

Figure 7.2
Pulse trains of decreasing frequency.

7.2 FOURIER TRANSFORMS

very similar in appearance to the sums by which Riemann integrals are defined. Letting $\lambda \to \infty$ and ignoring all technical difficulties, we formally obtain the expressions

$$\hat{U}(\phi) = \frac{1}{\sqrt{2\pi}} \int_{-\infty}^{\infty} u(t)e^{it\phi}\, dt, \qquad u(t) = \frac{1}{\sqrt{2\pi}} \int_{-\infty}^{\infty} U(\phi)e^{-it\phi}\, d\phi.$$

The similarity between the formulas for $\hat{U}$ and u is unmistakable; they are said to constitute a *Fourier transform pair*, and $u(t)$ is called the *Fourier transform* of $U(\phi)$. As in Section 7.1, the main problem is to discover under what circumstances the values $\hat{U}(\phi)$ and $U(\phi)$ coincide, as then $\hat{U}$ provides an *inversion* formula for the Fourier transform u. This has the effect of doubling the size of a given table of integrals, for if a closed form solution is known for the Fourier transform $u(t)$, one is also known for its inverse. The next theorem provides useful conditions under which $\hat{U}(\phi)$ and $U(\phi)$ agree, but is by no means the best theorem of this type.

Fourier Integral Theorem If $U(\phi)$ is piecewise smooth and $|U(\phi)|$ is integrable on $-\infty < \phi < \infty$, then

$$\text{PV } \hat{U}(\phi) = \text{PV } \frac{1}{\sqrt{2\pi}} \int_{-\infty}^{\infty} u(t)e^{it\phi}\, dt = \frac{1}{2}[U(\phi+0) + U(\phi-0)].$$

Proof Since $|U(\phi)|$ is integrable, the integral

$$\int_{-\infty}^{\infty} U(\phi)e^{it(\theta-\phi)}\, d\phi$$

converges uniformly with respect to t over any finite range. We may therefore integrate with respect to t over the interval $(-T, T)$, and invert the order of integration. Thus

$$\int_{-T}^{T} \int_{-\infty}^{\infty} U(\phi)e^{it(\theta-\phi)}\, d\phi\, dt = \int_{-\infty}^{\infty} U(\phi) \int_{-T}^{T} e^{it(\theta-\phi)}\, dt\, d\phi$$

$$= 2\int_{-\infty}^{\infty} U(\phi) \frac{\sin T(\theta-\phi)}{\theta-\phi}\, d\phi,$$

and $\Phi > |\theta| + 1$ (θ fixed) can be chosen such that

$$\int_{|\phi|>\Phi} |U(\phi)|\, d\phi < \frac{\varepsilon}{4}.$$

Then $\Phi - |\theta| > 1$, implying

$$\left| \int_{|\phi| > \Phi} U(\phi) \frac{\sin T(\theta - \phi)}{\theta - \phi} d\phi \right| \leq \int_{|\phi| > \Phi} |U(\phi)| \, d\phi < \frac{\varepsilon}{4}.$$

As in the first part of the proof of the Convergence Theorem for Fourier series in Section 7.1

$$\int_{\theta + \delta}^{\Phi} \frac{U(\phi)}{\theta - \phi} \sin T(\theta - \phi) \, d\phi \to 0 \quad \text{as} \quad T \to \infty, \tag{1}$$

and similarly for the integral on $[-\Phi, \theta - \delta]$, implying that for large T

$$\left| \int_{-T}^{T} \int_{-\infty}^{\infty} U(\phi) e^{it(\theta - \phi)} \, d\phi \, dt - 2 \int_{\theta - \delta}^{\theta + \delta} U(\phi) \frac{\sin T(\theta - \phi)}{\theta - \phi} d\phi \right| < \varepsilon.$$

But changing variables, we find

$$\int_{\theta - \delta}^{\theta + \delta} U(\phi) \frac{\sin T(\theta - \phi)}{\theta - \phi} d\phi = \int_{0}^{\delta} \frac{\sin T\phi}{\phi} [U(\theta + \phi) + U(\theta - \phi)] \, d\phi,$$

and it follows that

$$\text{PV } \hat{U}(\theta) = \lim_{T \to \infty} \frac{1}{\pi} \int_{0}^{\delta} \frac{\sin T\phi}{\phi} [U(\theta + \phi) + U(\theta - \phi)] \, d\phi. \tag{2}$$

Since the one-sided derivatives of U exist, the function

$$\frac{U(\theta + \phi) - U(\theta + 0) + U(\theta - \phi) - U(\theta - 0)}{\phi}$$

is piecewise smooth. By (1) the integral

$$\int_{0}^{\delta} \sin T\phi \left[\frac{U(\theta + \phi) + U(\theta - \phi) - U(\theta + 0) - U(\theta - 0)}{\phi} \right] d\phi \to 0$$

as $T \to \infty$, implying that

$$\text{PV } \hat{U}(\theta) = \lim_{T \to \infty} \frac{U(\theta + 0) + U(\theta - 0)}{\pi} \int_{0}^{\delta} \frac{\sin T\phi}{\phi} d\phi$$

$$= \frac{U(\theta + 0) + U(\theta - 0)}{\pi} \lim_{T \to \infty} \int_{0}^{T\delta} \frac{\sin \psi}{\psi} d\psi$$

$$= \frac{1}{2} [U(\theta + 0) + U(\theta - 0)]$$

by Dirichlet's Integral [Exercise 5, Section 2.3].

7.2 FOURIER TRANSFORMS

Example 1 Suppose $U(\phi) = e^{-|\phi|}$, then $|U(\phi)|$ is integrable and $U(\phi)$ is piecewise smooth. Its Fourier transform

$$u(t) = \frac{1}{\sqrt{2\pi}} \left[\int_{-\infty}^{0} e^{(-it+1)\phi} \, d\phi + \int_{0}^{\infty} e^{(-it-1)\phi} \, d\phi \right] = \frac{2}{\sqrt{2\pi}(1+t^2)}$$

satisfies

$$e^{-|\phi|} = \frac{1}{\pi} \int_{-\infty}^{\infty} \frac{e^{it\phi}}{1+t^2} \, dt = \frac{2}{\pi} \int_{0}^{\infty} \frac{\cos t\phi}{1+t^2} \, dt.$$

Compare this result with Example 1 in Section 4.2.

Example 2 Separating integrals as in the computation above we obtain

$$\int_{-\infty}^{\infty} e^{-y|t| - i(\phi - x)t} \, dt = \frac{2y}{(\phi - x)^2 + y^2},$$

transforming the Poisson Integral Formula for the upper half plane [Exercise 5, Section 6.2] into

$$U(z) = \frac{y}{\pi} \int_{-\infty}^{\infty} \frac{U(\phi)}{(\phi - x)^2 + y^2} \, d\phi = \frac{1}{2\pi} \int_{-\infty}^{\infty} \int_{-\infty}^{\infty} U(\phi) e^{-y|t| - i(\phi - x)t} \, dt \, d\phi.$$

Reversing the order of integration and letting $u(t)$ be the Fourier transform of $U(\phi)$ yields

$$U(z) = \frac{1}{\sqrt{2\pi}} \int_{-\infty}^{\infty} u(t) e^{-y|t| + ixt} \, dt = \left\{ \operatorname{Re} \frac{2}{\sqrt{2\pi}} \int_{0}^{\infty} u(t) e^{izt} \, dt \right\}, \quad (3)$$

since $\overline{u(t)} = u(-t)$ for real-valued functions $U(\phi)$ and

$$2 \operatorname{Re} u(t) e^{izt} = u(t) e^{izt} + \overline{u(t)} e^{-i\bar{z}t}$$
$$= u(t) e^{(-y+ix)t} + u(-t) e^{(y+ix)(-t)}.$$

Formula (3) is the analog for the half plane of the Fourier series expansion of the Poisson Integral Formula in the disk.

EXERCISES

1. Find the Fourier transforms of the functions:

(a) $\dfrac{b}{x^2 + b^2}$,

(b) $\dfrac{x}{x^2 + b^2}$,

(c) $\dfrac{x^2}{(x^2 + b^2)^2}$,

(d) $\dfrac{1}{x^4 + b^4}$.

(*Hint:* Use the contour integrals in Section 4.2.)

2. Find the Fourier transforms of the functions:

   (a) e^{-kx^2}, (b) xe^{-kx^2},

   (c) $\dfrac{1}{\sinh x}$, (d) $\dfrac{x}{\sinh x}$.

   (*Hint*: Use the integrals in Sections 2.2 and 4.5.)

3. Suppose $U(\phi) = 1/\sqrt{\phi}$ on $0 < \phi < \infty$ and vanishes on $-\infty < \phi \leq 0$. Find a function harmonic in the upper half plane having $U(\phi)$ as its boundary values.

4. Let the plate G shaped like the upper half plane have a temperature of $1°$ on the interval $[-1, 1]$ and $0°$ on the rest of the real axis. Find the temperature at every point of G.

5. Suppose $U = \hat{U}$ at almost every point in $(-\infty, \infty)$. Without worrying about convergence, show

$$\int_{-\infty}^{\infty} U(\phi)\overline{V(\phi)}\, d\phi = \int_{-\infty}^{\infty} u(t)\overline{v(t)}\, dt,$$

where u, v are the Fourier transforms of U, V, respectively. Then obtain *Parseval's Identity* for integrals

$$\int_{-\infty}^{\infty} |U(\phi)|^2\, d\phi = \int_{-\infty}^{\infty} |u(t)|^2\, dt.$$

6. Show

$$\int_{-\infty}^{\infty} e^{-2|\phi|}\, d\phi = \frac{2}{\pi} \int_{-\infty}^{\infty} \frac{dt}{(1+t^2)^2}.$$

   (*Hint*: Use Exercise 5.)

7.3 LAPLACE TRANSFORMS

Fourier transform methods often cannot be used in analyzing functions that are not absolutely integrable on $(-\infty, \infty)$. For example, the Heaviside function

$$H(\phi - a) = \begin{cases} 1, & \phi > a, \\ 0, & \phi < a, \end{cases}$$

7.3 LAPLACE TRANSFORMS

does not have a Fourier transform, as the integral

$$\int_a^\infty e^{-it\phi}\, d\phi$$

diverges. This is due to the fact the multiplier $e^{-it\phi}$ does not tend to zero as $\phi \to \infty$, leading us to try multipliers of the form $e^{-s\phi} = e^{-(q+it)\phi}$ which vanish when $q > 0$ as $\phi \to \infty$. The function

$$\mu(s) = \mathscr{L}_2\{U(\phi)\} = \int_{-\infty}^\infty U(\phi) e^{-s\phi}\, d\phi \tag{1}$$

is called the *two-sided Laplace transform* of the function $U(\phi)$. Writing $s = q + it$, equation (1) becomes

$$\int_{-\infty}^\infty U(\phi) e^{-q\phi} e^{-it\phi}\, d\phi,$$

which is the Fourier transform of the function $\sqrt{2\pi}\,U(\phi) e^{-q\phi}$.

Instead of developing an analysis of the two-sided Laplace transform, it is more convenient to write the integral in (1) in two parts

$$\int_{-\infty}^\infty U(\phi) e^{-s\phi}\, d\phi = \int_{-\infty}^0 U(\phi) e^{-s\phi}\, d\phi + \int_0^\infty U(\phi) e^{-s\phi}\, d\phi$$

$$= \int_0^\infty U(-\phi) e^{s\phi}\, d\phi + \int_0^\infty U(\phi) e^{-s\phi}\, d\phi.$$

Then a study of the properties of the integral

$$u(s) = \mathscr{L}\{U(\phi)\} = \int_0^\infty U(\phi) e^{-s\phi}\, d\phi, \tag{2}$$

called the (one-sided) *Laplace transform* of $U(\phi)$, enables us to investigate the behavior of the two-sided Laplace transform, since

$$\mathscr{L}_2\{U(\phi)\}(s) = \mathscr{L}\{U(-\phi)\}(-s) + \mathscr{L}\{U(\phi)\}(s).$$

The one-sided Laplace transform has many properties similar to those of power series. We shall prove the existence of a half plane of convergence analogous to the notion of the radius of convergence in Abel's Theorem. The two-sided Laplace transform will converge in a strip $a < \mathrm{Re}\, s < b\ (a \leq b)$ in analogy with the development of the Laurent series.

Theorem Suppose $U(\phi)$ is piecewise smooth and of *exponential order* (that is, there are real constants a and Φ such that $e^{-a\phi}|U(\phi)|$ is bounded for all $\phi > \Phi$.) Then $u(s)$ exists on Re $s > a$ (a *half plane of convergence*).

Proof Letting M be a finite bound for $e^{-a\phi}|U(\phi)|$ on $\phi > \Phi$ we have

$$\int_0^\infty |U(\phi)e^{-s\phi}|\,d\phi \leq \int_0^\Phi |U(\phi)e^{-s\phi}|\,d\phi + M\int_\Phi^\infty |e^{-(s-a)\phi}|\,d\phi.$$

But $U(\phi)$ is bounded on $[0, \Phi]$, since it is piecewise continuous, implying the first integral is finite, and

$$\int_\Phi^\infty |e^{-(s-a)\phi}|\,d\phi = \int_\Phi^\infty e^{-(\text{Re } s - a)\phi}\,d\phi = \frac{e^{-(\text{Re } s - a)\Phi}}{\text{Re } s - a} < \infty,$$

so the Laplace transform of $U(\phi)$ converges absolutely in Re $s > a$ (see Figure 7.3).

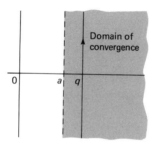

Figure 7.3
Half plane of convergence.

In fact, this last equation implies the Laplace transform converges uniformly in any closed set D lying entirely in the half plane of convergence, since

$$\left| \int_\phi^\infty U(\phi)e^{-s\phi}\,d\phi \right| \leq M \frac{e^{-(\text{Re } s - a)\phi}}{\text{Re } s - a},$$

for all $\phi \geq \Phi$, and ϕ can be chosen so as to make the right side of the equation above smaller than a preselected $\varepsilon > 0$, for all s in D.

For integers $n \geq 0$, let

$$u_n(s) = \int_0^n U(\phi)e^{-s\phi}\,d\phi.$$

7.3 LAPLACE TRANSFORMS

Then $u'_n(s)$ is a convergent integral in Re $s > a$, implying $u_n(s)$ is analytic in this domain. On any closed subset of this domain, the series of analytic functions

$$\sum_{n=0}^{\infty} [u_{n+1}(s) - u_n(s)] = \int_0^{\infty} U(\phi)e^{-s\phi}\, d\phi$$

is analytic, by Weierstrass's Theorem, implying the *Laplace transform u(s) is analytic in* Re $s > a$. Moreover, as we may then differentiate and integrate term-by-term, we have

$$\frac{d}{ds}\mathscr{L}\{U(\phi)\} = -\int_0^{\infty} \phi U(\phi)e^{-s\phi}\, d\phi = -\mathscr{L}\{\phi U(\phi)\}$$

and

$$\int_s^{\infty} \mathscr{L}\{U(\phi)\}\, ds = \int_0^{\infty} U(\phi)\left(\int_s^{\infty} e^{-s\phi}\, ds\right) d\phi$$

$$= \int_0^{\infty} \frac{U(\phi)}{\phi} e^{-s\phi}\, d\phi = \mathscr{L}\left\{\frac{U(\phi)}{\phi}\right\},$$

provided

$$\lim_{\phi \to 0^+} \frac{U(\phi)}{\phi} \text{ exists.}$$

Example 1

$$\mathscr{L}\{e^{-z\phi}\} = \frac{1}{s+z}, \qquad q = \text{Re } s > -\text{Re } z,$$

since in this domain

$$\int_0^{\infty} e^{-z\phi}e^{-s\phi}\, d\phi = \left.\frac{e^{-(s+z)\phi}}{-(s+z)}\right|_0^{\infty} = \frac{1}{s+z}.$$

Example 2

$$\mathscr{L}\{H(\phi - a)\} = \frac{e^{-as}}{s}, \qquad a \geq 0,$$

where $H(\phi - a) = 1$ when $\phi > a$ and vanishes elsewhere, because

$$\int_a^{\infty} e^{-s\phi}\, d\phi = \left.\frac{e^{-s\phi}}{-s}\right|_a^{\infty} = \frac{e^{-as}}{s}.$$

Example 3

$$\mathcal{L}\{\phi^n\} = \frac{n!}{s^{n+1}}, \quad n \geq 0.$$

By integrating by parts repeatedly we obtain

$$\int_0^\infty \phi^n e^{-s\phi}\, d\phi = \frac{\phi^n e^{-s\phi}}{-s}\bigg|_0^\infty + \frac{n}{s}\int_0^\infty \phi^{n-1} e^{-s\phi}\, d\phi$$

$$= \frac{n}{s}\left[\frac{\phi^{n-1} e^{-s\phi}}{-s}\bigg|_0^\infty + \frac{n-1}{s}\int_0^\infty \phi^{n-2} e^{-s\phi}\, d\phi\right]$$

$$\vdots$$

$$= \frac{n!}{s^n}\left[\int_0^\infty e^{-s\phi}\, d\phi\right] = \frac{n!}{s^{n+1}}.$$

Writing $s = q + it$, the Laplace transform becomes

$$u(s) = \int_0^\infty (U(\phi)e^{-q\phi})e^{-it\phi}\, d\phi, \quad q > a,$$

which is the Fourier transform of the function

$$P(\phi) = \begin{cases} \sqrt{2\pi}U(\phi)e^{-q\phi}, & \phi \geq 0, \\ 0, & \phi < 0. \end{cases}$$

Since $q > a$, $|P(\phi)|$ is integrable, so the Fourier Integral Theorem applies implying that at all points $\phi > 0$ of continuity of P we have

$$P(\phi) = \mathrm{PV}\,\hat{P}(\phi) = \frac{1}{\sqrt{2\pi}}\,\mathrm{PV}\int_{-\infty}^\infty u(q + it)e^{it\phi}\, dt$$

or

$$U(\phi) = \frac{1}{2\pi}\,\mathrm{PV}\int_{-\infty}^\infty u(q + it)e^{(q+it)\phi}\, dt,$$

$$= \frac{1}{2\pi i}\,\mathrm{PV}\int_{q-i\infty}^{q+i\infty} u(s)e^{s\phi}\, ds, \quad s = q + it, \quad \phi > 0. \tag{3}$$

This last equation is the *Inversion Theorem for Laplace Transforms*. We write $\mathcal{L}^{-1}\{u(s)\} = U(\phi)$ at all points $\phi > 0$ of continuity of U, and call U the *inverse* transform of u (see Figure 7.3). In particular, note that different continuous piecewise smooth functions of exponential order have different Laplace transforms.

7.3 LAPLACE TRANSFORMS

The methods of contour integration developed in Chapter 4 may be used in equation (3) to obtain the inverse of a given Laplace transform (see Example 5).

The development of an inversion formula for two-sided Laplace transforms is hampered by their nonuniqueness. The next example shows different functions may have identical two-sided Laplace transforms:

Example 4 Let

$$U_1(\phi) = \begin{cases} e^{-\phi} - e^{\phi}, & \phi > 0, \\ 0, & \phi \leq 0, \end{cases} \qquad U_2(\phi) = \begin{cases} e^{-\phi}, & \phi > 0, \\ e^{\phi}, & \phi \leq 0. \end{cases}$$

Then by Example 1 we have

$$\mathscr{L}\{U_1(\phi)\}(s) = \frac{1}{s+1} - \frac{1}{s-1}, \qquad \text{Re } s > 1,$$

$$\mathscr{L}\{U_1(-\phi)\}(-s) = 0,$$

while

$$\mathscr{L}\{U_2(\phi)\}(s) = \frac{1}{s+1}, \qquad \text{Re } s > -1,$$

$$\mathscr{L}\{U_2(-\phi)\}(-s) = \frac{-1}{s-1}, \qquad \text{Re } s < 1.$$

In both cases

$$\mathscr{L}_2\{U_k(\phi)\} = \frac{1}{s+1} - \frac{1}{s-1},$$

but the domain of convergence is Re $s > 1$ for $k = 1$, and $|\text{Re } s| < 1$ for $k = 2$.

Thus, any inversion formula for the two-sided Laplace transform must take the domain of convergence into consideration. One such special case is given by the Fourier transform, for which the domain of convergence contains the imaginary axis and $s = it$ in equations (1) and (3).

Example 5 Suppose we wish to find the inverse Laplace transform of a single-valued function $u(s)$, Re $s > a$. Suppose, in addition, that $u(s)$ vanishes as $s \to \infty$ in $\mathscr{M}$. Observe that for $s = q + Re^{i\theta}, q > a$,

$$|e^{s\phi}| = e^{q\phi + R\phi \cos \theta} \to 0$$

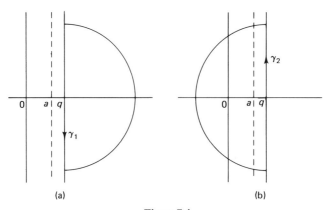

Figure 7.4

(a) $\phi < 0$. (b) $\phi > 0$.

as $R \to \infty$, provided $\phi \cos \theta < 0$. Thus, the contour integrals over the curves indicated in Figure 7.4

$$\frac{-1}{2\pi i} \int_{\gamma_1} u(s) e^{s\phi} \, ds, \qquad \phi < 0, \tag{4}$$

$$\frac{1}{2\pi i} \int_{\gamma_2} u(s) e^{s\phi} \, ds, \qquad \phi > 0, \tag{5}$$

converge to $\mathscr{L}^{-1}\{u(s)\}$ as $R \to \infty$. Since $u(s)$ is analytic in Re $s > a$, equation (4) vanishes by Cauchy's Theorem, implying that $U(\phi) = 0$, if $\phi < 0$. Finally, the Residue Theorem implies

$$U(\phi) = \sum_{\text{Re } s < q} \text{Res } u(s) e^{s\phi}, \qquad \phi > 0.$$

For example,

$$\mathscr{L}^{-1}\left\{\frac{1}{(s+1)(s+2)}\right\} = \text{Res}_{-1} \frac{e^{s\phi}}{(s+1)(s+2)}$$

$$+ \text{Res}_{-2} \frac{e^{s\phi}}{(s+1)(s+2)} = e^{-\phi} - e^{-2\phi}.$$

For multivalued functions $u(s)$, we must modify the curves γ_1, γ_2 so as to avoid crossing branch cuts (see Exercise 11).

7.3 LAPLACE TRANSFORMS

EXERCISES

1. Verify the following Laplace transforms and their domain of convergence:

   (a) $\mathscr{L}\{\cos z\phi\} = \dfrac{s}{s^2 + z^2}$, $\quad q > |\operatorname{Im} z|$,

   (b) $\mathscr{L}\{\cosh z\phi\} = \dfrac{s}{s^2 - z^2}$, $\quad q > |\operatorname{Re} z|$,

   (c) $\mathscr{L}\{\sin z\phi\} = \dfrac{z}{s^2 + z^2}$, $\quad q > |\operatorname{Im} z|$,

   (d) $\mathscr{L}\{\sinh z\phi\} = \dfrac{z}{s^2 - z^2}$, $\quad q > |\operatorname{Re} z|$.

2. Verify the following Laplace transforms:

   (a) $\mathscr{L}\{\phi \cos z\phi\} = \dfrac{s^2 - z^2}{(s^2 + z^2)^2}$, $\quad q > |\operatorname{Im} z|$,

   (b) $\mathscr{L}\{\phi \sin z\phi\} = \dfrac{2sz}{(s^2 + z^2)^2}$, $\quad q > |\operatorname{Im} z|$.

3. Without worrying about the domain of convergence, verify that:

   (a) $\mathscr{L}\left\{\dfrac{\sin z\phi}{\phi}\right\} = \tan^{-1}\dfrac{z}{s}$, $\quad q > |\operatorname{Im} z|$,

   (b) $\mathscr{L}\left\{\dfrac{\sinh z\phi}{\phi}\right\} = \dfrac{1}{2}\log\dfrac{s+z}{s-z}$, $\quad q > |\operatorname{Re} z|$.

4. Give an example of a piecewise smooth function that is not of exponential order.

5. Give an example of a function of exponential order that is not piecewise smooth.

6. Suppose $U(\phi)$ and $U'(\phi)$ are piecewise smooth functions of exponential order. Prove
$$\mathscr{L}\{U'(\phi)\} = s\mathscr{L}\{U(\phi)\} - U(0^+),$$
where
$$U(0^+) = \lim_{\phi \to 0^+} U(\phi).$$
Then show
$$\mathscr{L}\{\sin^2 z\phi\} = \dfrac{2z^2}{s(s^2 + 4z^2)}, \quad q > 2|\operatorname{Im} z|.$$

7. If $U(\phi)$ is piecewise smooth and of exponential order, show

$$\mathscr{L}\left\{\int_c^\phi U(\phi)\,d\phi\right\} = \frac{1}{s}\mathscr{L}\{U(\phi)\} + \frac{1}{s}\int_c^0 U(\phi)\,d\phi,$$

and use this to find the Laplace transform of the *sine integral*

$$\mathrm{Si}(\phi) = \int_0^\phi \frac{\sin\phi}{\phi}\,d\phi.$$

8. If $U(\phi)$ and $U'(\phi)$ are piecewise smooth and of exponential order, prove:
   (a) $\lim\limits_{s\to\infty} \mathscr{L}\{U(\phi)\} = 0$,
   (b) $\lim\limits_{s\to\infty} s\mathscr{L}\{U(\phi)\} = U(0^+)$,
   (c) $\lim\limits_{s\to 0^+} s\mathscr{L}\{U(\phi)\} = \lim\limits_{\phi\to\infty} U(\phi)$,

   provided the domain of convergence of $U'(\phi)$ includes the closed right half plane. Can the functions

$$\frac{s}{s-1},\quad \frac{1}{\sqrt{s}},\quad e^{s^{1/2}},$$

   be the Laplace transforms of function $U(\phi)$ which together with $U'(\phi)$ are piecewise smooth and of exponential order?

9. The unit "impulse" function, called the *delta function* $\delta(\phi-a)$, is loosely described as a "function" which is zero everywhere except at $\phi = a$ and having unit area under its graph

$$\int_{-\infty}^\infty \delta(\phi-a)\,d\phi = 1.$$

   Show that

$$H(\phi-a) = \int_{-\infty}^\phi \delta(\phi-a)\,d\phi,$$

   and if $a \geq 0$, prove

$$\mathscr{L}\{\delta(\phi-a)\} = e^{-as}.$$

10. Using the Inversion Formula, find the inverses of the following Laplace transforms in $\mathrm{Re}\,s > a$:

   (a) $\dfrac{1}{(s+a)^2}$,   (b) $\dfrac{1}{s^2+a^2}$,

7.4 PROPERTIES OF LAPLACE TRANSFORMS

(c) $\dfrac{1}{s(s^2 + a^2)}$, (d) $\dfrac{s}{(s^2 + a^2)^2}$,

(e) $\dfrac{s}{s^3 + a^3}$, (f) $\dfrac{1}{(s^3 + a^3)^2}$.

11. Find the inverse Laplace transform of the function

$$u(s) = \dfrac{1}{\sqrt{s}}, \quad \text{Re } s > 0.$$

(*Hint*: Use the contours indicated in Figure 7.5.)

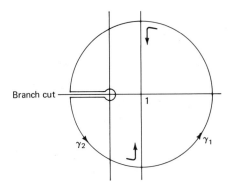

Figure 7.5

7.4 PROPERTIES OF LAPLACE TRANSFORMS

In this section we obtain several important facts about Laplace transforms, and give some applications of transform methods in the solution of differential equations. We shall assume the Laplace transform and its inverse transformation exist for the functions involved. It is clear, by separation of the integrand, that

$$\mathscr{L}\{aU(\phi) + bV(\phi)\} = a\mathscr{L}\{U(\phi)\} + b\mathscr{L}\{V(\phi)\},$$

so the Laplace transform is a *linear* operator. Clearly, so is the inverse operator $\mathscr{L}^{-1}$. Two useful results are the shifting theorems. Observe that

$$\int_0^\infty (U(\phi)e^{-w\phi})e^{-s\phi}\,d\phi = \int_0^\infty U(\phi)e^{-(s+w)\phi}\,d\phi,$$

yielding the *First Shifting Theorem*

$$\mathscr{L}\{U(\phi)e^{-w\phi}\}(s) = \mathscr{L}\{U(\phi)\}(s+w),$$

where the last expression means each s in $\mathscr{L}\{U\}$ is replaced by $s+w$. Also, by substitution, for $a \geq 0$

$$\int_0^\infty U(\phi)H(\phi-a)e^{-s\phi}\,d\phi = \int_a^\infty U(\phi)e^{-s\phi}\,d\phi$$
$$= e^{-as}\int_0^\infty U(\phi+a)e^{-s\phi}\,d\phi,$$

implying that

$$\mathscr{L}\{U(\phi)H(\phi-a)\} = e^{-as}\mathscr{L}\{U(\phi+a)\},$$

the *Second Shifting Theorem*, each ϕ in $U(\phi)$ being replaced by $\phi+a$.

Example 1 Evaluate $\mathscr{L}\{\sin z(\phi-a)H(\phi-a)\}$.

Using the Second Shifting Theorem, we find

$$\mathscr{L}\{\sin z(\phi-a)H(\phi-a)\} = e^{-as}\mathscr{L}\{\sin z\phi\} = \frac{ze^{-as}}{s^2+z^2}.$$

Example 2 Find the inverse of the Laplace transform $(s+w)^{-4}$.

By Example 3 of Section 7.3, we know

$$\mathscr{L}\left\{\frac{\phi^3}{3!}\right\} = \frac{1}{s^4},$$

so, using the First Shifting Theorem, we find

$$\frac{1}{(s+w)^4} = \mathscr{L}\left\{\frac{1}{6}e^{-w\phi}\phi^3\right\}.$$

Example 3 Solve the differential equation

$$U''(\phi) + 2wU'(\phi) + w^2U(\phi) = -\sin w\phi,$$

$$U(0) = 0, \qquad U'(0) = \frac{1}{2w}.$$

7.4 PROPERTIES OF LAPLACE TRANSFORMS

Using Exercises 6 and 1(c) of Section 7.3, we have

$$\mathscr{L}\{U''(\phi)\} = s\mathscr{L}\{U'(\phi)\} - \frac{1}{2w} = s^2\mathscr{L}\{U(\phi)\} - \frac{1}{2w},$$

$$\mathscr{L}\{U'(\phi)\} = s\mathscr{L}\{U(\phi)\},$$

so the Laplace transform of the differential equation becomes

$$(s^2 + 2ws + w^2)\mathscr{L}\{U(\phi)\} - \frac{1}{2w} = \frac{-w}{s^2 + w^2}.$$

Hence, by Example 1 and Exercise 1(a) of Section 7.3, we obtain

$$\mathscr{L}\{U(\phi)\} = \frac{s - w}{2w(s + w)(s^2 + w^2)}$$

$$= \frac{1}{2w^2}\left[\frac{s}{s^2 + w^2} - \frac{1}{s + w}\right]$$

$$= \frac{1}{2w^2}[\mathscr{L}\{\cos w\phi\} - \mathscr{L}\{e^{-w\phi}\}],$$

implying that

$$U(\phi) = \frac{\cos w\phi - e^{-w\phi}}{2w^2}.$$

This example demonstrates the power of transform methods, which allow us to replace the operations of differentiation and integration of the function $U(\phi)$ by simple algebraic operations on the transform $\mathscr{L}\{U(\phi)\}$.

The *convolution* of two functions $U(\phi)$ and $V(\phi)$ is the function

$$U * V(\phi) = \int_0^\phi U(t)V(\phi - t)\,dt.$$

Note that the roles of U and V can be interchanged without changing the value of the convolution. If the functions U, V are absolutely integrable on $(0, \infty)$ the convolution satisfies the identity

$$\mathscr{L}\{U * V\} = \mathscr{L}\{U\}\mathscr{L}\{V\},$$

that is, *the Laplace transform of the convolution is the product of the transforms of the functions.* The hypotheses above are sufficient to permit reversing the order of integration

$$\mathscr{L}\{U * V\} = \int_0^\infty \left[\int_0^\phi U(t)V(\phi - t)dt \right] e^{-s\phi} \, d\phi$$

$$= \int_0^\infty \left[\int_0^\infty U(t)V(\phi - t)H(\phi - t) \, dt \right] e^{-s\phi} \, d\phi$$

$$= \int_0^\infty U(t) \left[\int_0^\infty V(\phi - t)H(\phi - t)e^{-s\phi}d\phi \right] dt$$

$$= \int_0^\infty U(t)\mathscr{L}\{V(\phi - t)H(\phi - t)\} \, dt,$$

which by the Second Shifting Theorem yields

$$\mathscr{L}\{U * V\} = \mathscr{L}\{V\} \int_0^\infty U(t)e^{-ts} \, dt = \mathscr{L}\{U\}\mathscr{L}\{V\}.$$

The next two examples give an indication of the importance of the notion of convolution:

Example 4 To obtain a particular solution of the differential equation

$$U''(\phi) + 2wU'(\phi) + (w^2 + z^2)U(\phi) = V(\phi)$$

assume $U(0) = U'(0) = 0$, obtaining

$$[s^2 + 2ws + (w^2 + z^2)]\mathscr{L}\{U(\phi)\} = \mathscr{L}\{V(\phi)\}.$$

But by Exercise 1(c), Section 7.3, and the First Shifting Theorem, we have

$$\mathscr{L}\{e^{-w\phi} \sin z\phi\} = \frac{z}{(s + w)^2 + z^2}.$$

Thus

$$\mathscr{L}\{U(\phi)\} = \mathscr{L}\{z^{-1}e^{-w\phi} \sin z\phi\}\mathscr{L}\{V(\phi)\},$$

and we find

$$U(\phi) = \frac{1}{z} \int_0^\phi e^{-wt} \sin(zt)V(\phi - t) \, dt.$$

Example 5 Find the solution of the integral equation

$$\int_0^\phi (\phi - t)^{\alpha - 1} U(t) \, dt = V(\phi), \qquad 0 < \alpha < 1,$$

where $V(\phi)$ is piecewise smooth and of exponential order for $\phi \geq 0$.

7.4 PROPERTIES OF LAPLACE TRANSFORMS

Let $u(s) = \mathscr{L}\{U(\phi)\}$, $v(s) = \mathscr{L}\{V(\phi)\}$. Observe that by Exercise 8, Section 4.5

$$\Gamma(\beta) = \int_0^\infty e^{-t} t^{\beta-1}\, dt = s^\beta \int_0^\infty \phi^{\beta-1} e^{-\phi s}\, d\phi,$$

with β, Re $s > 0$. Hence

$$\mathscr{L}\{\phi^{\beta-1}\} = \Gamma(\beta) s^{-\beta}, \qquad \beta > 0.$$

Applying the Laplace transform to the given convolution yields

$$v(s) = \mathscr{L}\{\phi^{\alpha-1} \ast U(\phi)\} = \Gamma(\alpha) s^{-\alpha} u(s).$$

Thus

$$u(s) = \frac{s^\alpha}{\Gamma(\alpha)} v(s) = s \cdot \frac{s^{\alpha-1}}{\Gamma(\alpha)} v(s).$$

But

$$\mathscr{L}\{\phi^{-\alpha}\} = \Gamma(1-\alpha) s^{\alpha-1} = \frac{\pi}{\sin \alpha \pi} \frac{s^{\alpha-1}}{\Gamma(\alpha)},$$

since $\Gamma(\alpha)\Gamma(1-\alpha) = \pi/\sin \alpha\pi$ (see Exercise 2, Section 8.3), so

$$u(s) = \frac{\sin \alpha \pi}{\pi} \cdot s\mathscr{L}\{\phi^{-\alpha} \ast V(\phi)\}.$$

The convolution

$$F(\phi) = \int_0^\phi (\phi - t)^{-\alpha} V(t)\, dt$$

and its derivative satisfy the hypothesis of Exercise 6, Section 7.3, with $F(0^+) = 0$, hence $\mathscr{L}\{F'(\phi)\} = s\mathscr{L}\{F(\phi)\}$ and

$$U(\phi) = \frac{\sin \alpha \pi}{\pi} \frac{d}{d\phi} \int_0^\phi (\phi - t)^{-\alpha} V(t)\, dt.$$

These last two examples motivate the important concept of a transfer function. Many physical systems may be thought of as devices which transform a given input function V into an output function U. Assuming all initial conditions are zero when $\phi = 0$ and taking the Laplace transform of the equations describing the system, we obtain the expression

$$\mathscr{L}\{U(\phi)\} = \frac{\mathscr{L}\{V(\phi)\}}{Z(s)},$$

where $Z(s)^{-1}$, the *transfer function*, is independent of V. Let U_H be the output function when $V(\phi) = H(\phi)$. Then, using Example 2, Section 7.3, we find

$$\mathscr{L}\{U\} = \frac{s\mathscr{L}\{V\}}{sZ(s)} = s\mathscr{L}\{U_H\}\mathscr{L}\{V\} = s\mathscr{L}\{U_H * V\}.$$

So, by Exercise 6, Section 7.3,

$$U(\phi) = (U_H * V)'(\phi) = \int_0^\phi U_H(t)V'(\phi - t)\,dt + U_H(\phi)V(0). \qquad (1)$$

By commuting $\mathscr{L}\{U_H\}$ and $\mathscr{L}\{V\}$ above, it also follows that

$$U(\phi) = (V * U_H)'(\phi) = \int_0^\phi V(t)U_H'(\phi - t)\,dt, \qquad (2)$$

since the initial conditions imply $U_H(0) = 0$. Equations (1) and (2) are called *Duhamel's Formulas*, and express the response of a system to an input function $V(\phi)$ in terms of the experimentally accessible response to the Heaviside function.

EXERCISES

Assume the existence and inversion of the transforms in Exercises 2–10.
1. Using the Shifting Theorems, evaluate:
   (a) $\mathscr{L}\{e^{-w\phi}\cos z\phi\}$,   (b) $\mathscr{L}\{e^{-w\phi}\sin z\phi\}$,
   (c) $\mathscr{L}\{e^{-w\phi}\phi^n\}$,   (d) $\mathscr{L}\{e^{-z\phi}H(\phi - a)\}$.

2. What are the domains of convergence of the two Shifting Theorems?
3. Show

$$U(\phi) = \frac{d}{d\phi}\mathscr{L}^{-1}\left\{\frac{1}{s}\mathscr{L}\{U(\phi)\}\right\},$$

provided the right side exists for all $\phi > 0$. Apply the above in finding

$$\mathscr{L}^{-1}\left\{\frac{s}{(s-1)(s+2)}\right\}.$$

(*Hint*: Use Exercises 6 and 8 of Section 7.3.)
4. Prove

$$U(\phi) = \int_0^\phi \mathscr{L}^{-1}\{s\mathscr{L}\{U(\phi)\}\}\,d\phi,$$

7.4 PROPERTIES OF LAPLACE TRANSFORMS

and apply it in finding

$$\mathscr{L}^{-1}\left\{\frac{1}{s(s^2+4)}\right\}.$$

(*Hint*: Use Exercise 7, Section 7.3.)

5. Prove convolution is distributive, commutative, and associative:
   (a) $U * (V_1 + V_2) = U * V_1 + U * V_2$,
   (b) $U * V = V * U$,
   (c) $(U * V) * W = U * (V * W)$.

(*Hint*: Assume the fact (which we have not proved) that different *continuous* functions have different Laplace transforms. Show the convolution is a continuous function.)

6. Solve the differential equation

$$U''(\phi) + 3U'(\phi) + 2U(\phi) = \sin \phi, \qquad U(0) = U'(0) = 0,$$

using Laplace transforms. (*Hint*: Apply Exercise 6, Section 7.3, and expand in partial fractions.)

7. Solve the differential equation

$$U''(\phi) + U(\phi) = \phi \sin \phi, \qquad U(0) = 0, \quad U'(0) = 1,$$

using Laplace transforms.

8. Find the Laplace transform of the general solution of the differential equation

$$\phi U''(\phi) + U'(\phi) + \phi U(\phi) = 0.$$

9. Find $\mathscr{L}\{J_0(\phi)\}$ and compare with the solution of Exercise 8. (*Hint*: Expand $J_0(\phi)$ as a Taylor series.)

10. Find a solution of the partial differential equation

$$U_{tt} = a^2 U_{xx},$$

having boundary and initial conditions

$$U(x, 0) = 0, \quad U(b, t) = 0, \qquad U(0, t) = \sin \frac{a\pi t}{b}, \qquad U_t(x, 0) = 0,$$

where a and b are fixed constants. (This problem arises, for example, in studying the vibrations of a string.) (*Hint*: Let $u(x, s) = \mathscr{L}\{U(x, t)\}$, assume the order of the processes of differentiation and integration may be reversed in obtaining $\mathscr{L}\{U_{xx}\}$ and $\mathscr{L}\{U_{tt}\}$, and solve the resulting ordinary differential equation. The Shifting Theorems will be employed in obtaining the solution.)

NOTES

Section 7.1 An example of a continuous function whose Fourier series diverges at the rational numbers in $[0, 2\pi]$ can be found in [J, p. 546]. A comprehensive discussion of the convergence problem can be found in [Hf]. Term-by-term integrations and differentiations may be performed on Fourier series, obtaining the Fourier series of the indefinite integral or derivative, if the functions concerned are piecewise smooth.

Section 7.2 Alternative definitions for Fourier transforms are frequently encountered. All such definitions are equivalent up to a rotation and magnification by $\sqrt{2\pi}$, hence constitute a personal prejudice of the user.

Section 7.3 Tables of Laplace transforms may be found in many mathematical handbooks. One such table is in [M, pp. 428–434].

Section 7.4 For a proof of the uniqueness of the Laplace transform for continuous functions see [M, p. 412].

Chapter 8 | ASYMPTOTIC EXPANSIONS

8.1 DEFINITIONS AND PROPERTIES

By Laurent's Theorem, a function $f(z)$, analytic at $z = \infty$, can be represented by a power series in $1/z$ converging absolutely in $|z| > R$. Letting $S_n(z)$ denote the nth partial sum of the series, we know that not only

$$\lim_{n \to \infty} |f(z) - S_n(z)| = 0, \qquad |z| > R,$$

but also

$$\lim_{|z| \to \infty} |f(z) - S_n(z)| = 0, \qquad n = 0, 1, 2, \ldots.$$

If $f(z)$ is not analytic at $z = \infty$, no such convergent power series exists. However, it is often possible to generate a power series in $1/z$ such that *for each n and a restricted range of* arg z

$$\lim_{|z| \to \infty} \left\{ z^n \left[\frac{f(z)}{g(z)} - \sum_{k=0}^{n} \frac{a_k}{z^k} \right] \right\} = 0,$$

where $g(z)$ is a function whose behavior for large values of z is known. In this case we write

$$f(z) \sim g(z) \sum_{k=0}^{\infty} \frac{a_k}{z^k},$$

and call the right side an *asymptotic expansion* of $f(z)$. For large values of z, an asymptotic expansion approximates the function, since in the restricted range of arg z we have

$$\left| f(z) - g(z) \sum_{k=0}^{n} \frac{a_k}{z^k} \right| < \varepsilon \frac{|g(z)|}{|z|^n}, \qquad \varepsilon = \varepsilon(n, z).$$

Often, this approximation is highly accurate and may be used to calculate the value of $f(z)$, but unlike convergent series, taking additional terms needs not improve the accuracy, because ε depends on n. Typically, ε tends to infinity as $n \to \infty$.

Example 1 The *exponential integral* given by

$$\text{Ei}(z) = \int_z^\infty \frac{e^{-\zeta}}{\zeta} d\zeta \qquad (1)$$

converges along any curve γ joining z to ∞ in the domain $|\arg z| < \frac{1}{2}\pi$. To prove this fact consider Figure 8.1. The integral of $e^{-\zeta}/\zeta$ along the bound-

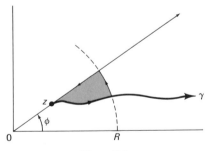

Figure 8.1

8.1 DEFINITIONS AND PROPERTIES

ary of the shaded area vanishes by Cauchy's Theorem. But the integrals along the circular and radial arcs of this contour are bounded by

$$\int_{-\pi/2}^{\pi/2} |e^{-Re^{i\theta}}| \, d\theta \leq 2e^{-R} \int_0^{\pi/2} e^{2R\theta/\pi} \, d\theta = \frac{\pi}{R}(1 - e^{-R}) \to 0 \quad \text{as} \quad R \to \infty$$

[since $h(\theta) = \cos\theta + (2\theta/\pi) - 1$ vanishes at $\theta = 0, \pi/2$ and satisfies $h''(\theta) < 0$ in $(0, \pi/2)$], and

$$\int_{|z|}^{\infty} \frac{|e^{-re^{i\phi}}|}{r} \, dr \leq \frac{1}{|z|} \int_{|z|}^{\infty} e^{-r\cos\phi} \, dr$$

$$= \frac{e^{-|z|\cos\phi}}{|z|\cos\phi}, \qquad \phi = \arg z.$$

In particular, the values of the integrals along the radial line from z to ∞ and along γ are identical, explaining the notation in equation (1).

Integrating by parts, we obtain

$$\text{Ei}(z) = \frac{e^{-z}}{z} - \int_z^{\infty} \frac{e^{-\zeta}}{\zeta^2} \, d\zeta = \frac{e^{-z}}{z} - \frac{e^{-z}}{z^2} + 2 \int_z^{\infty} \frac{e^{-\zeta}}{\zeta^2} \, d\zeta,$$

and after n repetitions

$$\text{Ei}(z) = e^{-z}\left(\frac{1}{z} - \frac{1}{z^2} + \frac{2}{z^3} - \cdots + \frac{(-1)^{n-1}(n-1)!}{z^n}\right) + (-1)^n n! \int_z^{\infty} \frac{e^{-\zeta}}{\zeta^{n+1}} \, d\zeta.$$

But the last term, called the *remainder*, is dominated by

$$n! \int_{|z|}^{\infty} \frac{e^{-r\cos\phi}}{r^{n+1}} \, dr < \frac{n! e^{-|z|\cos\phi}}{|z|^{n+1}\cos\phi}, \qquad \phi = \arg z, \tag{2}$$

hence

$$\left| z^n \left(\frac{\text{Ei}(z)}{e^{-z}} - \sum_{k=1}^{n} \frac{(-1)^{k-1}(k-1)!}{z^k}\right) \right| < \frac{n!}{|z|\cos\phi} \to 0$$

as $|z| \to \infty$. Thus

$$\text{Ei}(z) \sim e^{-z}\left(\sum_{k=1}^{\infty} \frac{(-1)^{k-1}(k-1)!}{z^k}\right), \qquad \text{Re } z > 0.$$

Note the series diverges everywhere in $\text{Re } z > 0$, and the smallest value of the remainder is obtained by taking n equal to the integral part of $|z|$, since the right side of equation (2) may be written in the form

$$\frac{1}{|z|} \cdot \frac{2}{|z|} \cdots \frac{n}{|z|} \cdot \frac{e^{-|z|\cos\phi}}{|z|\cos\phi}.$$

For $n = z = 10$, the error of the approximation is less than

$$\frac{10!e^{-10}}{10^{11}} \approx 1.6475 \times 10^{-9}.$$

Figure 8.2 contains the graphs of the first four partial sums of the asymptotic expansion of $e^z\text{Ei}(z)$ along the interval $[1, 7]$ of the real axis. The number of terms in the partial sum is indicated adjacent to its graph, and the graph of the function $e^x\text{Ei}(x)$ is represented by the broken line in the figure.

Observe the partial sum with n terms is the best approximation at $x = n$,

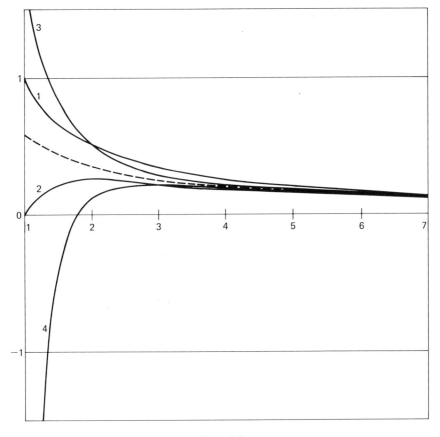

Figure 8.2
Graph of $e^x\text{Ei}(x)$, represented by the broken line, and its approximations.

8.1 DEFINITIONS AND PROPERTIES

$n = 1, 2, 3, 4$. The partial sums with an odd number of terms exceed the function $e^x \text{Ei}(x)$, which in turn exceeds those with an even number of terms.

We now prove some properties of asymptotic expansions. Given $f(z) \sim g(z)$ $(a_0 + a_1/z + \cdots)$ we also have $f(z)/g(z) \sim a_0 + a_1/z + \cdots$, so we need only work with expansions of the latter form.

If $f(z) \sim a_0 + a_1/z + \cdots$, in $\alpha < \arg z < \beta$, this representation is unique, for if not

$$\lim_{|z| \to \infty} z^n \left[f(z) - a_0 - \frac{a_1}{z} - \cdots - \frac{a_n}{z^n} \right]$$

$$= 0$$

$$= \lim_{|z| \to \infty} z^n \left[f(z) - b_0 - \frac{b_1}{z} - \cdots - \frac{b_n}{z^n} \right], \qquad \alpha < \arg z < \beta,$$

implying that

$$\lim_{|z| \to \infty} z^n \left[(a_0 - b_0) + \frac{a_1 - b_1}{z} + \cdots + \frac{a_n - b_n}{z^n} \right] = 0, \qquad \alpha < \arg z < \beta,$$

and therefore $a_0 = b_0, a_1 = b_1, \ldots, a_n = b_n$. However, the same asymptotic expansion may be valid for different functions. Observe that

$$\lim_{|z| \to \infty} z^n e^{-z} = 0, \qquad |\arg z| < \frac{\pi}{4},$$

implying $e^{-z} \sim 0$ in $|\arg z| < \pi/4$. So if $f(z) \sim a_0 + a_1/z + \cdots$ in $|\arg z| < \pi/4$, then $f(z) + e^{-z} \sim a_0 + a_1/z + \cdots$ in $|\arg z| < \pi/4$.

Integrating an asymptotic expansion yields the asymptotic expansion of the integral. Letting $f(z) \sim a_0 + a_1/z + \cdots$, in $\alpha < \arg z < \beta$, and $\varepsilon > 0$, there is an $R = R(n)$ such that

$$\left| z^n \left(f(z) - a_0 - \frac{a_1}{z} - \cdots - \frac{a_n}{z^n} \right) \right| < \varepsilon,$$

whenever $|z| > R$, $\alpha < \arg z < \beta$. For any z_0 in this domain

$$\left| \int_{z_0}^{\infty} \left(f(z) - a_0 - \frac{a_1}{z} - \cdots - \frac{a_n}{z^n} \right) dz \right| < \varepsilon \int_{|z_0|}^{\infty} \frac{dr}{r^n} = \frac{\varepsilon}{(n-1)|z_0|^{n-1}},$$

hence we have

$$\left| z_0^{n-1} \left[\int_{z_0}^{\infty} \left(f(z) - a_0 - \frac{a_1}{z} \right) dz - \int_{z_0}^{\infty} \left(\frac{a_2}{z^2} + \cdots + \frac{a_n}{z^n} \right) dz \right] \right| < \frac{\varepsilon}{n-1},$$

and thus

$$\int_\infty^z \left(f(z) - a_0 - \frac{a_1}{z}\right) dz \sim \frac{a_2}{z} + \frac{a_3}{2z^2} + \frac{a_4}{3z^3} + \cdots, \qquad \alpha < \arg z < \beta.$$

From the integration and uniqueness properties it follows that *if $f'(z)$ has an asymptotic expansion, it is obtained by term-by-term differentiation of the asymptotic expansion of $f(z)$.*

EXERCISES

1. Show that if $f(z)$ is analytic in $R < |z| < \infty$ and $f(z) \sim a_0 + a_1/z + \cdots$ for all values of arg z, then $f(z)$ is analytic at $z = \infty$. (*Hint*: Show $f(z) - (a_0 z^n + \cdots + a_n)$ has a removable singularity at ∞.)
2. If $f(z) \sim a_0 + a_1/z + \cdots$ and $g(z) \sim b_0 + b_1/z + \cdots$, show:

   (a) $Af(z) + Bg(z) \sim (Aa_0 + Bb_0) + (Aa_1 + Bb_1)/z + \cdots$,
   (b) $f(z)g(z) \sim a_0 b_0 + (a_1 b_0 + a_0 b_1)/z + (a_2 b_0 + a_1 b_1 + a_0 b_2)/z^2 + \cdots$.

   (*Hint*: Write (b) as three terms which tend to 0.)
3. Show Ei(z) converges on the set $|\arg z| \leq \pi/2$, $|z| > 0$.
4. Does Ei(z) converge in the sector $|\arg z| < \pi$?
5. Verify the asymptotic expansions in $0 \leq \arg z \leq \pi$, $|z| > 0$, for the integrals

$$\int_z^\infty \frac{\cos z}{z} dz \sim (\cos z)\left(\frac{1}{z^2} - \frac{3!}{z^4} + \frac{5!}{z^6} - \cdots\right) - (\sin z)\left(\frac{1}{z} - \frac{2!}{z^3} + \frac{4!}{z^5} - \cdots\right),$$

$$\int_z^\infty \frac{\sin z}{z} dz \sim (\sin z)\left(\frac{1}{z^2} - \frac{3!}{z^4} + \frac{5!}{z^6} - \cdots\right) + (\cos z)\left(\frac{1}{z} - \frac{2!}{z^3} + \frac{4!}{z^5} - \cdots\right).$$

For $z = x$, use Dirichlet's Integral [Exercise 5, Section 2.3] to compute the sine integral,

$$\text{Si}(x) = \int_0^x \frac{\sin x}{x} dx,$$

for $x = 10$. Show your error is less than 3.6288×10^{-5} and compare your answer with the table value Si(10) = 1.65834. (*Hint*: Substitute $-iz$ for z in the exponential integral.)
6. Integrating by parts, obtain an asymptotic expansion for the function

$$F(z) = \int_1^\infty e^{-zt} \frac{1}{\sqrt{t}} dt, \qquad t \text{ real}, \quad \text{Re } z > 0.$$

Compute $F(10)$ and estimate your error.

8.2 METHOD OF STEEPEST DESCENT

7. Consider the differential equation

$$w'' + \left(1 - \frac{k^2}{z^2}\right)w = 0.$$

For large values of $|z|$, it seems reasonable to neglect the term involving k^2/z^2, obtaining the equation $w'' + w = 0$. This equation has the two independent solutions, $\cos z$ and $\sin z$, suggesting

$$w \sim (\cos z)\left(a_0 + \frac{a_1}{z} + \frac{a_2}{z^2} + \cdots\right) + (\sin z)\left(b_0 + \frac{b_1}{z} + \frac{b_2}{z^2} + \cdots\right).$$

By substituting the asymptotic expansion into the differential equation, determine the constants a_j and b_j, $j = 1, 2$, in terms of a_0 and b_0.

8.2 METHOD OF STEEPEST DESCENT

The method to be explained in the next two sections is very useful in determining the asymptotic behavior of integral representations of analytic functions. We shall need the following result:

Theorem (Watson) Let $f(x)$ be analytic and bounded on a domain containing the real axis, and t be a positive real number. Then

$$\int_{-\infty}^{\infty} e^{-t^2 x^2/2} f(x)\, dx \sim \sqrt{2\pi}\left(\frac{a_0}{t} + \frac{a_2}{t^3} + \frac{1\cdot 3}{t^5}a_4 + \frac{1\cdot 3\cdot 5}{t^7}a_6 + \cdots\right), \quad (1)$$

where the a_j's are the coefficients of the Maclaurin series of $f(z)$.

Proof Note that the function

$$g(z) = \frac{f(z) - (a_0 + a_1 z + \cdots + a_{2n-1} z^{2n-1})}{z^{2n}} = \frac{R_{2n}(z)}{z^{2n}} \quad (2)$$

is bounded (say, by M) on the real axis, since it tends to a finite limit as $z \to 0$ and $f(z)$ is bounded on the real axis. Repeatedly integrating by parts we have, for integers $n \geq 0$,

$$\int_{-\infty}^{\infty} e^{-t^2 x^2/2} x^{2n}\, dx$$

$$= \frac{2n-1}{t^2}\int_{-\infty}^{\infty} e^{-t^2 x^2/2} x^{2n-2}\, dx$$

$$\vdots$$

$$= \frac{1\cdot 3 \cdots (2n-1)}{t^{2n}}\int_{-\infty}^{\infty} e^{-t^2 x^2/2}\, dx = \sqrt{2\pi}\,\frac{1\cdot 3 \cdots (2n-1)}{t^{2n+1}}$$

by letting $\sqrt{2}u = tx$ and using equation (2), Section 2.2. But (2) above implies that

$$\left| \int_{-\infty}^{\infty} e^{-t^2 x^2/2} [f(x) - (a_0 + a_1 x + \cdots + a_{2n-1} x^{2n-1})] \, dx \right|$$

$$\leq M \int_{-\infty}^{\infty} e^{-t^2 x^2/2} x^{2n} \, dx,$$

and as the integrands with odd powers of x are odd functions, their integrals vanish and we have

$$\left| \int_{-\infty}^{\infty} e^{-t^2 x^2/2} f(x) \, dx - \sqrt{2\pi} \left(\frac{a_0}{t} + \frac{a_2}{t^3} + \frac{1 \cdot 3}{t^5} + \cdots + \frac{1 \cdot 3 \cdots (2n-3)}{t^{2n-1}} a_{2n-2} \right) \right|$$

$$\leq \sqrt{2\pi} \, M \frac{1 \cdot 3 \cdots (2n-1)}{t^{2n+1}}.$$

Multiplying both sides by $t^{2n}/\sqrt{2\pi}$ and letting $t \to \infty$ yields the result.

The *Method of Steepest Descent* (or the *Saddle Point Method*) is applied in the approximate evaluation of integrals of the form

$$F(t) = \int_{\gamma} e^{tf(z)} g(z) \, dz, \qquad t > 0, \tag{3}$$

where γ is a pwd curve in $\mathscr{C}$ and $f(z)$, $g(z)$ are analytic in a domain containing γ. To avoid unnecessary complications, we shall assume γ has endpoints α and β and $f(z)$ and $g(z)$ are entire.

Writing $f = u + iv$, observe that if t is large, $\cos tv$ and $\sin tv$ may oscillate at high frequency even for small displacements along γ. In addition, if e^{tu} is large, the resulting integrations of large values of opposite sign complicates any attempt to approximate the value of the integral. However, if the pwd curve γ' also joins α to β, then by Cauchy's Theorem

$$\int_{\gamma} e^{tf(z)} g(z) \, dz = \int_{\gamma'} e^{tf(z)} g(z) \, dz,$$

since the integrand is entire. Hopefully, the difficulties mentioned above may be avoided by a judicious choice of the γ' path of integration. We must seek a pwd curve joining α to β along which e^{tu} is small whenever v changes, and any large values of e^{tu} are taken on segments along which v is constant.

The procedure now is to draw the level lines $u(z) = $ constant and $v(z) = $ constant. These curves are normal to one another as was shown in Section 5.5.

8.2 METHOD OF STEEPEST DESCENT

The domains where u is positive are called *hills*, and those where u is negative are called *valleys*, the level lines of u serving to indicate the "elevation." We later show the level lines of v indicate the direction of maximum change in the value of u. If α and β lie in the same valley, the path must never go outside the valley, but if they are in different valleys we must find a path which avoids the hills as far as possible. In some instances the hills are unavoidable. For example, suppose

$$u + iv = f(z) = e^z = e^x \cos y + ie^x \sin y.$$

The lines $u = 0$ are horizontal lines through $(k + \tfrac{1}{2})\pi i$, $k = 0, \pm 1, \pm 2, \ldots$. The remaining level lines of u are indicated by solid lines in Figure 8.3. Observe that the level lines of v, indicated by dashed lines are identical in appearance, but 90° out of phase. The hills of the relief map of u are shaded. In this example it is not possible to go from $1 - i\pi$ to $1 + i\pi$ without crossing a hill. However, the e^x term makes the hills right of the imaginary axis higher

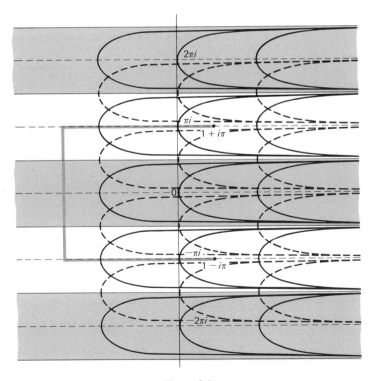

Figure 8.3

and the valleys deeper, while it flattens out those to the left of the imaginary axis. Thus our selection of a path γ should cross the strip $|y| < \pi/2$ well to the left of the imaginary axis to keep u small, as most of the contribution to the integral occurs in this band.

Often a mountain pass connects two valleys as shown in Figure 8.4. Such a pass is called a *saddle*, and the point z_0 is called a *saddle point*. It is apparent

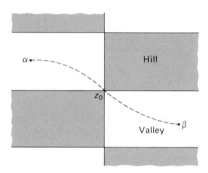

Figure 8.4

Saddle point.

there will be an advantage in selecting a path through z_0 which descends from the pass as quickly as possible. Thus, the path should follow the line of *steepest descent*, and hence the name of the method. The direction of such a path coincides with the direction of the maximal directional derivative at each point, that is, that value θ maximizing

$$u_s(\theta) = u_x \cos \theta + u_y \sin \theta. \qquad (4)$$

Differentiating equation (4) with respect to θ and setting it equal to zero, we have by the Cauchy–Riemann equations

$$0 = -u_x \sin \theta + u_y \cos \theta = -(v_x \cos \theta + v_y \sin \theta) = -v_s(\theta),$$

implying v does not change along this path. Hence the *line of steepest descent* coincides with a level line of $v(z)$. This, of course, is precisely what we were hoping for.

A word of warning is necessary in that $g(z)$ has been ignored in the analysis above. If $g(z)$ has very large values in a neighborhood of the saddle point, it might be necessary to select a path of integration that does not pass through

8.2 METHOD OF STEEPEST DESCENT

the saddle point. Whenever possible, the path should stay on a level line of $v(z)$. As such a situation might yield a substantial contribution to the value of the integral (3), it is usually necessary to obtain bounds for the integral on this path.

Singularities of the functions $f(z)$ and $g(z)$ introduce additional difficulties. In this case, after selecting the path γ' we must take into account the residues of the singularities of $f(z)$ and $g(z)$ inside $\gamma - \gamma'$. Then the integral along γ equals that along γ' together with the residues of the singularities.

Example 1 If $f(z) = z^2$ and $g(z) = 1$, find the line of steepest descent joining $-i$ to i.

Since $u = x^2 - y^2$ and $v = 2xy$ the level lines coincide with the family of hyperbolas indicated in Figure 8.5. The shaded domains are those on which $u > 0$, thus 0 is a saddle point. Since the imaginary axis is part of the level

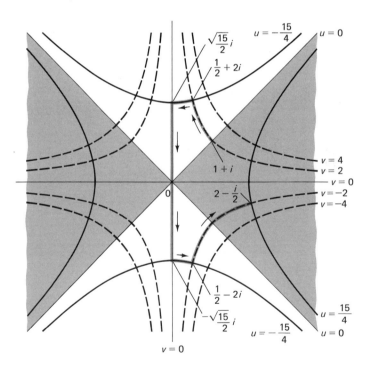

Figure 8.5

A family of hyperbolas.

lines $v = 0$, the line of steepest descent is along the imaginary axis. If the endpoints are $1 + i$ and $2 - i/2$, select the path indicated by a railroad track in Figure 8.5 passing along level lines of u and v through the points $\frac{1}{2} + 2i$, $\sqrt{15i/2}$, $-\sqrt{15i/2}$.

EXERCISES

1. Let $f(z) = -i \sin z$ and $g(z) = 1$. Find the line of steepest descent joining $-\pi + i$ to $\pi + i$.
2. Let $f(z) = z^3$ and $g(z) = 1$. Find the line of steepest descent joining -1 to $1 + i$.
3. Show there is only one line of steepest descent from a saddle point into a valley.
4. Show that the lines of steepest descent terminate only at singularities of $f(z)$ or at infinity.
5. Show

$$F(t) = \int_{-\infty}^{\infty} e^{-t^2 x^2/2} \cos x \, dx \sim \frac{\sqrt{2\pi}}{t}\left(1 - \frac{1}{2t^2} + \frac{1}{2^2 2! t^4} - \frac{1}{2^3 3! t^6} + \cdots\right),$$

and compute $F(10)$. Compare your answer with the correct value obtained in Exercise 2, Section 2.2]. (In fact the first two terms yields surprisingly accurate answers.)
6. Show

$$F(t) = \int_{-\infty}^{\infty} e^{-t^2 x^2/2} \cos(x^2) \, dx$$

$$\sim \frac{\sqrt{2\pi}}{t}\left(1 - \frac{4!}{2^2(2!)^2 t^4} + \frac{8!}{2^4(4!)^2 t^8} - \frac{12!}{2^6(6!)^2 t^{12}} + \cdots\right),$$

and compute $F(10)$.

8.3 CONTINUATION

We now complete our discussion of the Method of Steepest Descent by analyzing the contribution each saddle point makes to the integral

$$F(t) = \int_{\gamma} e^{tf(z)} g(z) \, dz.$$

Once this amount has been determined, the asymptotic value of the integral is obtained by summing these saddle point contributions.

8.3 CONTINUATION

Expanding $f(z)$ in a Taylor series centered on the saddle point z_0, we have

$$f(z) = f(z_0) + f'(z_0)(z - z_0) + \frac{f''(z_0)}{2!}(z - z_0)^2 + \cdots.$$

But since z_0 is a saddle point, $u_x(z_0)$ and $u_y(z_0)$ both vanish, hence by the Cauchy–Riemann equations

$$f'(z_0) = f_x(z_0) = u_x(z_0) + iv_x(z_0) = 0.$$

The first term of the asymptotic expansion is now easy to obtain if $f''(z_0) \neq 0$. We have, by Section 3.1,

$$f(z) - f(z_0) = (z - z_0)^2 f_2(z),$$

with $f_2(z)$ analytic in a neighborhood of z_0 and tending to $f''(z_0)/2$ as $z \to z_0$. The integral on the part γ' of the path lying in this neighborhood is approximately equal to

$$e^{tf(z_0)} \int_{\gamma'} e^{tf''(z_0)(z-z_0)^2/2} g(z)\, dz.$$

Substituting $-\zeta^2 = f''(z_0)(z - z_0)^2$ we obtain

$$\frac{e^{tf(z_0)}}{\sqrt{-f''(z_0)}} \int_{\gamma''} e^{-t\zeta^2/2} g\!\left(z_0 + \frac{\zeta}{\sqrt{-f''(z_0)}}\right) d\zeta,$$

where γ'', the image of γ', is part of the line of steepest descent. But the exponential is similar to that in Example 1 of the last section, except that it is of opposite sign. This reverses the location of the hills and valleys, implying the shaded domains are now valleys, so the line of steepest descent is along the real axis. As the exponential approaches 0 rapidly with increasing $|x|$, no significant error is made by assuming the limits of integration are $-\infty$ and $+\infty$. Thus the saddle points' contribution is approximately

$$\frac{e^{tf(z_0)}}{\sqrt{-f''(z_0)}} \int_{-\infty}^{\infty} e^{-tx^2/2} g\!\left(z_0 + \frac{x}{\sqrt{-f''(z_0)}}\right) dx \sim \frac{e^{tf(z_0)}}{\sqrt{-f''(z_0)}} \cdot \sqrt{2\pi}\, \frac{g(z_0)}{\sqrt{t}},$$

by the theorem in the previous section, since $g(z_0)$ is the first coefficient of the Maclaurin series of the function

$$G(\zeta) = g\!\left(z_0 + \frac{\zeta}{\sqrt{-f''(z_0)}}\right).$$

This last quantity is the contribution the saddle point z_0 makes to the integral $F(t)$. If only one saddle point lies on γ, we have

$$F(t) \sim e^{tf(z_0)} g(z_0) \sqrt{\frac{2\pi}{-tf''(z_0)}}. \tag{1}$$

Again, we emphasize, that in certain situations it might be necessary to include contributions other than those caused by saddle points, in order to get a good approximation.

We apply the result above in analyzing the Gamma function:

Example 1 The real Gamma function can be defined by the integral

$$\Gamma(t+1) = \int_0^\infty e^{-x} x^t \, dx, \qquad t > -1.$$

(Compare this with Exercise 8, Section 4.5.) For large $t > 0$, if we let $x = ts$ and $dx = t\, ds$, we obtain

$$\Gamma(t+1) = t^{t+1} \int_0^\infty e^{t(-s + \ln s)} \, ds.$$

Setting $f(z) = -z + \ln z$ and $g(z) = 1$, the level lines of $u(z) = \operatorname{Re} f = -x + \ln |z|$ are the solid lines indicated in Figure 8.6.

The level lines of u are easily found by a graphical method due to *Maxwell*, applicable whenever u can be written as the sum of two functions for which the level lines are known. If $u = u_1 + u_2$, then the points of intersection of the level lines $u_1 = a$ and $u_2 = b$ are points on the level line $u = a + b$, as are the points of intersection of the level lines $u_1 = a + k$ and $u_2 = b - k$. In this fashion it is possible to generate a large number of points on a fixed level line of u. In our case, let $u_1 = -x$ and $u_2 = \ln |z|$; their level lines are the dashed lines in Figure 8.6.

The shaded domain corresponds to the points z, where $u(z) > -1$, so $z = 1$ is a saddle point. The positive real axis is the level line $v(z) = -y + i \arg z = 0$. Since 1 is the only saddle point, equation (1) implies

$$\Gamma(t+1) = t^{t+1} \int_0^\infty e^{t(-s + \ln s)} \, ds \sim t^{t+1} e^{-t} \sqrt{\frac{2\pi}{t}},$$

as $f''(z) = -z^{-2}$ and $f(1) = f''(1) = -1$. Thus we obtain *Stirling's Formula*

$$\Gamma(t+1) \sim \sqrt{2\pi}\, t^{t+1/2} e^{-t}.$$

8.3 CONTINUATION

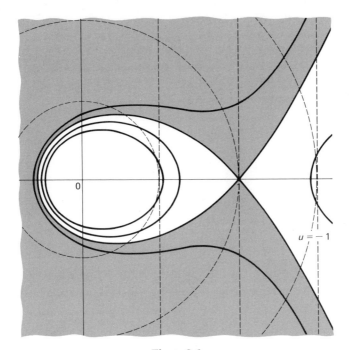

Figure 8.6
Level lines of $u = -x + \log|z|$.

EXERCISES

1. The complex Gamma function was defined in Exercise 8, Section 3.5, by

$$\Gamma(z+1) = \int_0^\infty e^{-s} s^z \, ds, \quad 0 < s < \infty, \quad \operatorname{Re} z > -1.$$

Extend the proof in Example 1 and show

$$\Gamma(z+1) \sim \sqrt{2\pi} \, z^{z+1/2} e^{-z}.$$

2. Prove $\Gamma(t)\Gamma(1-t) = \pi/\sin \pi t$, $0 < t < 1$. (*Hint*: Set $x = u^2$ in the definition of $\Gamma(t)$, $x = v^2$ for $\Gamma(1-t)$ and convert to polar coordinates.)
3. Show Bessel's function $J_n(z)$ defined in Exercise 5, Section 3.3, by

$$J_n(x) = \frac{1}{2\pi i} \int_{|\zeta|=1} \frac{e^{(x/2)(\zeta - 1/\zeta)}}{\zeta^{n+1}} \, d\zeta, \quad \text{for} \quad z = x > 0,$$

8 ASYMPTOTIC EXPANSIONS

Figure 8.7

may also be obtained by integration over the contour $\gamma - \gamma^*$, where γ^* is the reflection across the real axis of the curve γ indicated in Figure 8.7. The integrals

$$H_n^{(1)}(x) = \frac{1}{\pi i} \int_\gamma \frac{e^{(x/2)(\zeta - 1/\zeta)}}{\zeta^{n+1}} \, d\zeta,$$

$$H_n^{(2)}(x) = \frac{1}{\pi i} \int_{-\gamma^*} \frac{e^{(x/2)(\zeta - 1/\zeta)}}{\zeta^{n+1}} \, d\zeta,$$

are called (real) *Hankel functions* and satisfy $2J_n(x) = H_n^{(1)}(x) + H_n^{(2)}(x)$.

4. Show by the Method of Steepest Descent that

$$H_n^{(1)}(x) \sim \sqrt{\frac{2}{\pi x}} \, e^{i(x - n\pi/2 - \pi/4)},$$

$$H_n^{(2)}(x) \sim \sqrt{\frac{2}{\pi x}} \, e^{-i(x - n\pi/2 - \pi/4)},$$

hence

$$J_n(x) \sim \sqrt{\frac{2}{\pi x}} \cos\left(x - \frac{n\pi}{2} - \frac{\pi}{4}\right).$$

(*Hint*: The orientation of the curve determines the sign of the approximation. This occurs at the γ'' stage of the development of the method.)

5. Show that the applications of the Method of Steepest Descent to Exercise 4 can be extended to complex values z, obtaining

$$H_n^{(1)}(z) \sim \sqrt{\frac{2}{\pi z}} \, e^{i(z - n\pi/2 - \pi/4)},$$

$$H_n^{(2)}(z) \sim \sqrt{\frac{2}{\pi z}} \, e^{-i(z - n\pi/2 - \pi/4)},$$

and

$$J_n(z) \sim \sqrt{\frac{2}{\pi z}} \cos\left(z - \frac{n\pi}{2} - \frac{\pi}{4}\right).$$

NOTES

Section 8.1 Tables for the sine integral and exponential integral may be found in most mathematical handbooks.

Sections 8.2 and 8.3 The Method of Stationary Phase is largely equivalent to the Method of Steepest Descent. A brief discussion may be found in [CKP, pp. 272–274]. A discussion concerning the procedure required to obtain the saddle point contribution when $f''(z_0) = 0$ can be found in [CKP, p. 264]. Numerous applications of these techniques to physical problems are to be found in the succeeding pages.

APPENDIX

A.1 TABLE OF CONFORMAL MAPPINGS

z-plane	Mapping function	w-plane
	$w = i\dfrac{1-z}{1+z}$	
	$\left(\dfrac{w-1}{w+1}\right)^2 = i\dfrac{z-1}{z+1}$	
	$w = \text{Log}\, z$	
	$w = \text{Log}\, i\left(\dfrac{1-z}{1+z}\right)$	
	$z = \dfrac{|\alpha|}{\alpha}\left(\dfrac{w-\alpha}{1-\bar{\alpha}w}\right)$	

APPENDIX

z-plane	Mapping function	w-plane
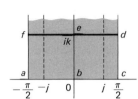	$w = \sin z$ Solid line: $$\left(\frac{u}{\cosh k}\right)^2 + \left(\frac{v}{\sinh k}\right)^2 = 1$$ Dashed lines: $$\left(\frac{u}{\sin j}\right)^2 - \left(\frac{v}{\cos j}\right)^2 = 1$$	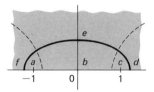
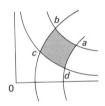	$w = z^2$ $\{xy = c\} \to \{v = 2c\}$ $\{x^2 - y^2 = c\} \to \{u = c\}$	
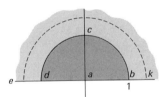	$w = z + \dfrac{1}{z}$ Dashed line: $$\left(\frac{ku}{k^2+1}\right)^2 + \left(\frac{kv}{k^2-1}\right)^2 = 1$$	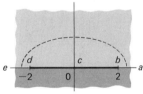
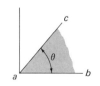	$w = z^{\pi/\theta}$	

REFERENCES

[A]   Ahlfors, L. V. *Complex Analysis*, 2nd ed. McGraw-Hill, New York, 1966.
[B]   Buck, R. C. *Advanced Calculus*. McGraw-Hill, New York, 1965.
[CKP] Carrier, G. F., Krook, M., and Pearson, C. E. *Functions of a Complex Variable*. McGraw-Hill, New York, 1966.
[H]   Hille, E. *Analytic Function Theory*, Vols. I and II. Ginn (Blaisdell), Boston, Massachusetts, 1959.
[Hf] Hoffman, K. *Banach Spaces of Analytic Functions*. Prentice-Hall, Englewood Cliffs, New Jersey, 1962.
[Ho] Hormander, L. *An Introduction to Complex Analysis in Several Variables*. Van Nostrand-Reinhold, Princeton, New Jersey, 1966.
[J]   James, R. C. *Advanced Calculus*. Wadsworth, Belmont, California, 1966.
[Kn] Knopp, K. *Theory of Functions*, Parts I and II. Dover, New York, 1947.

REFERENCES

[Ko] Kober, H. *Dictionary of Conformal Representations*, 2nd ed. Dover, New York, 1957.

[M] Moretti, G. *Functions of a Complex Variable*. Prentice-Hall, Englewood Cliffs, New Jersey, 1964.

[R] Rothe, R., Ollenderf, F., and Pohlhausen, K. *Theory of Functions*. Dover, New York, 1961.

[S] Saks, S. *Theory of the Integral*, 2nd rev. ed. Dover, New York, 1964.

[Sp] Springer, G. *Introduction to Riemann Surfaces*. Addison-Wesley, Reading, Massachusetts, 1957.

[T] Titchmarsh, E. C. *The Theory of Functions*, 2nd ed. Oxford Univ. Press, London and New York, 1939.

[V] Veech, W. A. *A Second Course in Complex Analysis*. Benjamin, New York, 1967.

[W] Whyburn, G. T. *Topological Analysis*, rev. ed. Princeton Univ. Press, Princeton, New Jersey, 1964.

INDEX

Page numbers indicate where the entry is defined.

A

Abel's theorem, 67
Absolute convergence, 59
Absolute value, 2
Accumulation point, 63
Analytic function, 12
   continuation, 80, 82
Antiderivative, 38
Arc, 29
   opposite, 29
   simple, 29
Arc length, 30
Argument, 2
Argument principle, 111
Asymptotic expansion, 192

B

Bessel's function, 76, 206
Boundary, 8
Bounded, 8
Branch, 20, 85
   cut, 20
   logarithmic, 85
   point, 20
   principal, 2, 22, 23

C

Cauchy estimate, 53
Cauchy–Goursat theorem, 34
Cauchy integral formula, 46
Cauchy principal value, 99
Cauchy–Riemann equations, 13
Cauchy theorem, 40, 50
Charge, 140, 160
Circles of Appolonius, 156
Circulation, 135
Closed, 8
Closed curve, 29
Closure, 10
Complement, 8
Complex conjugate, 4
Complex number, 2
Complex plane, 2
Complex potential, 136
Condenser, 141
Conductor, 141
Conformal mapping, 117
Connected, 8
   multiply, 42
   simply, 9
Converges, 58, 99
Convolution, 185

D

De Moivre's theorem, 7, 21
Delta function, 182
Derivative, 12
Differential arc, 29
Dipole, 158, 160

Dirichlet's integral, 45, 101
Dirichlet's problem, 147
Diverges, 59
Domain, 9
Doublet, 158
Duhamel's formulas, 188

E

Electrostatic field, 140
Elements, 80
Elliptic integral, 131
Endpoints, 29
Entire function, 12
Equipotential line, 137, 141
Essential singularity, 77
Exponential, 19
  integral, 192
  order, 176
Exterior, 8
  angle, 129

F

Flow
  heat, 138
  irrotational, 135
  steady, 134
Fluid flow, 134
  incompressible, 134
Force lines, 141
Fourier coefficients, 164
Fourier integral theorem, 171
Fourier law, 139
Fourier series, 164
Fourier transform, 171
Fresnel's integrals, 45
Function(s)
  analytic, 12
  Bessel, 76, 206
  complex valued, 10
  continuous, 11
  force, 141
  gamma, 84, 111, 204
  global analytic, 85

Hankel, 206
harmonic, 146
Heaviside, 174
impulse, 103
meromorphic, 77
potential, 136
regular, 27
stream, 136
transfer, 188
Fundamental theorem
  algebra, 55
  integration, 40

G

Gamma function, 84, 111, 204
Gauss mean value theorem, 53
Global, 85, 118
Gradient, 139
Green's theorem, 45

H

Hadamard's formula, 66
Half-plane of convergence, 176
Harmonic conjugate, 146
Harmonic function, 146
Harnack's inequality, 151
Heat flow, 138
Heaviside function, 174
Hill, 199
Holomorphic, 27

I

Image, 10
Imaginary, pure, 2
Imaginary axis, 2
Imaginary, part, 2
Imaginary, unit, 2
Impulse function, 103
Incompressible, 134
Infinity, 5
Inside, 7, 29

Intensity, 155
Interior, 8
Inversion, 120
Irrotational, 135
Isolated, 62
  singularity, 76
Isotherm, 139

J

Jacobian, 119
Jordan arc, 29
Jordan curve, 29
Joukowski profile, 160

L

Laguerre polynomial, 53
Laplace equation, 145
  one-sided transform, 175
  two-sided transform, 175
Laurent series, 72
Laurent theorem, 72
Legendre polynomial, 53
Lemniscate, 155
L'Hospital's theorem, 71
Limit, 11
Line of steepest descent, 200
Linear fractional transformation, 119
Liouville's theorem, 54
Logarithm, 22
Looman–Menchoff, 27

M

Maclaurin series, 61
Magnification, 120
Mapping, conformal, 117
Maximum principle, 55, 146
Maxwell's graphical method, 204
Mean value theorem, 146
Menchoff, 15
Meromorphic, 77

Method
  stationary phase, 207
  steepest descent, 198
Minimum principle, 56, 146
Mittag–Leffler, 88
Modulus, 2
Moment, dipole, 156
Monodromy theorem, 83
Monogenic, 27
Morera's theorem, 51, 54
Multiplet, 161
Multiply connected, 42

N

Natural boundary, 84
Neighborhood, 7
Number
  complex, 2
  real, 1

O

One-to-one, 10
  locally, 118
Onto, 10
Open, 8
Order
  branch point, 85
  exponential, 176
  multiplet, 161
  pole, 76
  zero, 61
Origin, 2
Outside, 29

P

Parseval's identity, 165, 174
Picard's theorem, 78, 88
Piecewise differentiable, 29
  smooth, 166
Poisson's integral formula, 147, 149
Poles, 76
Positive orientation, 29

Potential, 136
  field, 141
Principal value, 2, 22, 23, 99
Pringsheim's theorem, 88

R

Radius of convergence, 65
Real axis, 2
Real number, 1
Real part, 2
Regular point, 85
Removable singularity, 76
Residue theorem, 90
Riemann mapping theorem, 128
Riemann sphere, 6
Riemann surface, 20, 84
Riemann theorem, 48
Rotation, 120
Rouche's theorem, 112

S

Saddle point, 200
Schwarz–Christoffel formula, 128
Schwarz formula, 151
Schwarz lemma, 56
Shifting theorems, 184
Simply connected, 9
Sine integral, 182, 196
Singularities, 76, 85
Sinks, 134, 154
Source, 134, 154
  strength, 154
Stagnation point, 137
Stationary phase, 207
Steepest descent
  line, 200
  method, 200

Stereographic projection, 6
Stirling's formula, 204
Streamlines, 136, 139
Sum, 59
Symmetric, 125
Symmetry principle, 125

T

Taylor series, 59
Taylor theorem, 59
Temperature, 139
Thermal conductivity, 139
Three-circles theorem, 57
Transfer function, 188
Translation, 120
Triangle inequality, 5

U

Unbounded, 8
Uniform convergence, 64

V

Valley, 199
Variation, 112
Velocity vector, 134
Vortex, 154
  source, 155

W

Watson's theorem, 197
Weierstrass–Casorati theorem, 78
Weierstrass product, 88
Weierstrass theorem, 64